263
Topics in Current Chemistry

Topics in Current Chemistry

Recently Published and Forthcoming Volumes

Supramolecular Chirality
Volume Editors: Crego-Calama, M., Reinhoudt, D. N.
Vol. 265, 2006

Radicals in Synthesis II
Complex Molecules
Volume Editor: Gansäuer, A.
Vol. 264, 2006

Radicals in Synthesis I
Methods and Mechanisms
Volume Editor: Gansäuer, A.
Vol. 263, 2006

Molecular Machines
Volume Editor: Kelly, T. R.
Vol. 262, 2006

Immobilisation of DNA on Chips II
Volume Editor: Wittmann, C.
Vol. 261, 2005

Immobilisation of DNA on Chips I
Volume Editor: Wittmann, C.
Vol. 260, 2005

Prebiotic Chemistry
From Simple Amphiphiles to Protocell Models
Volume Editor: Walde, P.
Vol. 259, 2005

Supramolecular Dye Chemistry
Volume Editor: Würthner, F.
Vol. 258, 2005

Molecular Wires
From Design to Properties
Volume Editor: De Cola, L.
Vol. 257, 2005

Low Molecular Mass Gelators
Design, Self-Assembly, Function
Volume Editor: Fages, F.
Vol. 256, 2005

Anion Sensing
Volume Editor: Stibor, I.
Vol. 255, 2005

Organic Solid State Reactions
Volume Editor: Toda, F.
Vol. 254, 2005

DNA Binders and Related Subjects
Volume Editors: Waring, M. J., Chaires, J. B.
Vol. 253, 2005

Contrast Agents III
Volume Editor: Krause, W.
Vol. 252, 2005

Chalcogenocarboxylic Acid Derivatives
Volume Editor: Kato, S.
Vol. 251, 2005

New Aspects in Phosphorus Chemistry V
Volume Editor: Majoral, J.-P.
Vol. 250, 2005

Templates in Chemistry II
Volume Editors: Schalley, C. A., Vögtle, F., Dötz, K. H.
Vol. 249, 2005

Templates in Chemistry I
Volume Editors: Schalley, C. A., Vögtle, F., Dötz, K. H.
Vol. 248, 2004

Collagen
Volume Editors: Brinckmann, J., Notbohm, H., Müller, P. K.
Vol. 247, 2005

Radicals in Synthesis I

Methods and Mechanisms

Volume Editor: Andreas Gansäuer

With contributions by

A. Barchuk · F. Brebion · D. Crich · K. Daasbjerg · V. Darmency
A. Gansäuer · M. Gerenkamp · S. Grimme · C. Mück-Lichtenfeld
P. Renaud · M. P. Sibi · D.-H. Suk · H. Svith · J. Zimmerman · H. Zipse

Springer

The series *Topics in Current Chemistry* presents critical reviews of the present and future trends in modern chemical research. The scope of coverage includes all areas of chemical science including the interfaces with related disciplines such as biology, medicine and materials science. The goal of each thematic volume is to give the nonspecialist reader, whether at the university or in industry, a comprehensive overview of an area where new insights are emerging that are of interest to a larger scientific audience.

As a rule, contributions are specially commissioned. The editors and publishers will, however, always be pleased to receive suggestions and supplementary information. Papers are accepted for *Topics in Current Chemistry* in English.

In references *Topics in Current Chemistry* is abbreviated Top Curr Chem and is cited as a journal.

Visit the TCC content at springerlink.com

ISSN 0340-1022
ISBN-10 3-540-31329-X Springer Berlin Heidelberg New York
ISBN-13 978-3-540-31329-8 Springer Berlin Heidelberg New York
DOI 10.1007/11420187

Springer is a part of Springer Science+Business Media

springer.com

Printed in Germany

Cover design: *Design & Production* GmbH, Heidelberg
Typesetting and Production: LE-TEX Jelonek, Schmidt & Vöckler GbR, Leipzig

Printed on acid-free paper 02/3100 YL – 5 4 3 2 1 0

Volume Editor

Prof. Dr. Andreas Gansäuer

Kekulé-Institut für Organische
Chemie und Biochemie
Gerhard-Domagk-Straße 1
53121 Bonn, Germany
andreas.gansaeuer@uni-bonn.de

Editorial Board

Topics in Current Chemistry
Also Available Electronically

For all customers who have a standing order to Topics in Current Chemistry, we offer the electronic version via SpringerLink free of charge. Please contact your librarian who can receive a password or free access to the full articles by registering at:

springerlink.com

If you do not have a subscription, you can still view the tables of contents of the volumes and the abstract of each article by going to the SpringerLink Homepage, clicking on "Browse by Online Libraries", then "Chemical Sciences", and finally choose Topics in Current Chemistry.

You will find information about the

- Editorial Board
- Aims and Scope
- Instructions for Authors
- Sample Contribution

at springer.com using the search function.

Preface

"I didn't think that radical chemistry could be so mild and selective," is the nicer version of comments one often hears after seminars. What is the underlying reason for the misconception? Probably that radical transformations often seem counterintuitive to those brought up with classical retrosynthetic schemes. As a result, the use of radicals is considered by many synthetic chemists as a last resort only to be used when other more traditional methods have failed. Additionally, radical reactions are usually regarded as being unselective and involving toxic reagents.

This is, of course, false; such a conservative approach neglects the mild, selective, and original solutions available through using radical chemistry for demanding synthetic problems. Moreover, a solid physical organic understanding of the mechanism behind most radical reactions has now been established. This basis serves us well in predicting many results as well as in developing novel reactions. In short, radical chemistry has developed with amazing speed from a laboratory curiosity into an integral, predictable, and highly productive part of organic chemistry. This account is meant to further spread this point of view.

The first volume (*Methods and Mechanisms*) concentrates on the mechanistic aspects of radical chemistry and the development of novel methods, while the second volume (*Complex Molecules*) focuses on the use of radicals in synthetic applications. While such traditional separation (novel methods are increasingly aimed at preparing complex molecules and the synthesis of complex molecules requires careful planning) may seem a little outdated at the beginning of the 21st century, it is nevertheless employed for the sake of convenience.

The chapters, written by leading experts, provide state-of-the-art reviews of exciting and pertinent topics of current research in radical chemistry. These include a discussion of computed data concerning radical stabilities and their evaluation, the surprising chemistry of radical cations, modern concepts and reagents for enantioselective radical chemistry, the mechanistic aspects of epoxide opening via electron transfer, the evolution of ecologically benign and efficient tin-free radical reactions, the attractive novel reagents and radical traps for unusual cyclizations, the exciting possibilities of xanthate derived radical processes, the emerging field of radical chemistry on solid supports, the recent development of highly versatile radical tandem reactions, the mild and selective derivatization of amino acids and sugars through the use of radicals, and the increasing use of Cp_2TiCl-catalyzed and -mediated radical reactions in natural product synthesis.

Of course not all of the exciting recent developments in radical chemistry can be covered in depth in just two books. It is therefore planned to expand this series in the near future. I offer my apologies to the authors left out this time and ask them to contribute next time!

Hopefully this book will meet the challenge of convincing a large number of scientists of the benefits of radical chemistry and spark novel developments in the fields of new radical methodology and the application of radical reactions in the synthesis of complex molecules.

Bonn, February 2006 Andreas Gansäuer

Contents

Contents of Volume 264

Radicals in Synthesis II

Volume Editor: Andreas Gansäuer

ISBN: 3-540-31325-7

Top Curr Chem (2006) 263: 1–38
DOI 10.1007/128_024

Published online: 20 December 2005

Generation of Alkene Radical Cations by Heterolysis of β-Substituted Radicals: Mechanism, Stereochemistry, and Applications in Synthesis

David Crich (✉) · Franck Brebion · Dae-Hwan Suk

Department of Chemistry, University of Illinois at Chicago, 845 West Taylor Street, Chicago, IL 60607-7061, USA
Dcrich@uic.edu

Abstract The experimental basis for the formation of alkene radical cations by the heterolysis of alkyl radicals bearing leaving groups at the β position is reviewed, and a general mechanism involving contact alkene radical cation/anion pairs is presented for both fragmentation reactions and rearrangements. The available kinetic data for both fragmentations and migrations are summarized. The β-(acyloxy)alkyl and β-(phosphatoxy)alkyl radical rearrangements, previously viewed as concerted shifts, are reinterpreted in terms of the general mechanism with extremely rapid collapse of the intermediate contact alkene radical cation/anion pair. The reactions of alkene radical cations in the confines of the contact ion pair are reviewed, including radical cyclizations, nucleophilic attack, and tandem nucleophilic attack/radical cyclization processes. Stereochemical memory effects arising from the order within the contact alkene radical cation/anion pair are discussed at the level of both diastereoselectivity and enantioselectivity.

Keywords Alkene radical cations · Ion pairs · Kinetics · Stereochemical memory effects · Tandem reactions

1 Introduction

Alkene radical cations are charged, open-shell reactive intermediates formally arising by the one-electron oxidation of a $C=C$ π bond (Scheme 1). These cations display facets of both free radical and cation chemistry, but it is the combination of the two that renders them particularly fascinating, and which confers novel patterns of reactivity on them. (For an insightful discourse on the need to include both the radical and ionic components of radical ions when considering reactivity, see [1].) Classically, this group of reactive intermediates has been generated from alkenes, essentially according to Scheme 1, using a variety of different oxidizing protocols including chemical one-electron oxidants, anodic oxidation, direct photochemical electron ejection, and photostimulated one-electron oxidants. This relative ease of generation has resulted in a wealth of studies of alkene radical cation reactivity, which has been covered before in this series and in a number of other books, reviews, and recent articles [2–31]. However direct, this classical method of alkene radical generation imposes severe limitations on functional group compatibility unless the alkene to be oxidized is somewhat electron rich. It is only within the last decade that an alternative method for alkene radical generation, not relying on the one-electron oxidation of alkenes, has been developed and begun to be applied in synthesis. This method relies on the expulsion of leaving groups from the β position of free radicals (Scheme 2), which may themselves be generated under a wide variety of con-

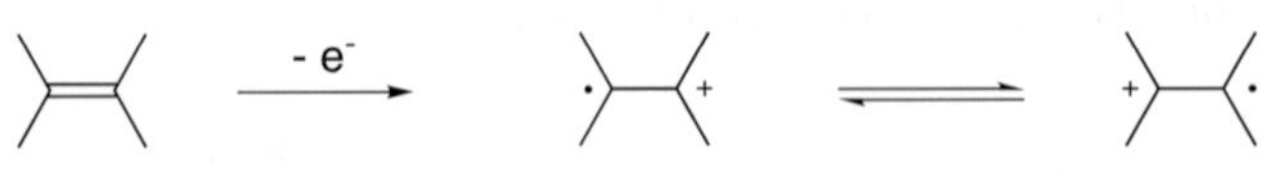

Scheme 1 Generation of alkene radical cations from alkenes

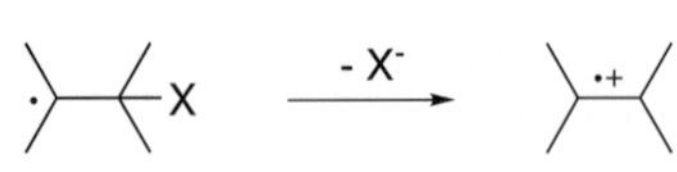

Scheme 2 Generation of alkene radical cations by the expulsion of a leaving group

ditions, including reductive ones. The genesis of this new method and its continuing evolution form the subject matter of this chapter.

2
Background and Historical Perspectives

The group of Norman and coworkers was the first to postulate the expulsion of a leaving group from the β position of an alkyl radical in their electron spin resonance (ESR) study of the β-acetoxy-α-methoxyethyl radical [32]. These researchers generated this radical under Fenton conditions from the corresponding alkane but only observed the spectrum of a rearranged radical. It was suggested that this rearranged radical arose by an initial heterolytic fragmentation to give an alkene radical cation, followed by nucleophilic trapping by the solvent, water. Working with the same β-acetoxy-α-methoxyethyl radical but generated under pulse radiolytic conditions, Schulte-Frohlinde and coworkers observed the same rearranged radical as that seen by the Norman group as well as a regioisomer (Scheme 3) [33]. This regioisomer was seen to rearrange under the acidic conditions of the experiment to give the obviously thermodynamic radical detected by the Norman group. As in the initial formation of the two regioisomeric products, the interconversion was seen as proceeding via an alkene radical cation. The thermodynamic preference for the β-hydroxy-β-methoxyethyl radical arises from the anomeric interaction between the two C – O bonds.

The expulsion of phosphate groups from the β position of alkyl radicals, and particularly α-alkoxyalkyl radicals, has long been recognized to be an important phenomenon in the cleavage of oligonucleotides (Scheme 4) [34–36]. The cleavage of DNA C4′ radicals has been extensively studied in recent years, and was the subject of several review articles [37–45], before achieving prominence as a means of hole injection into DNA bases for the study of electron transfer along the oligonucleotide backbone [46, 47].

In parallel with the development of the heterolysis of β-substituted alkyl radicals, a rearrangement reaction was observed and extensively studied in organic solvents. This rearrangement was first noted for β-(acyloxy)alkyl radicals (Scheme 5) by Surzur et al. [48] and, later, for β-(phosphatoxy)alkyl radicals by the Crich and Giese groups [49, 50].

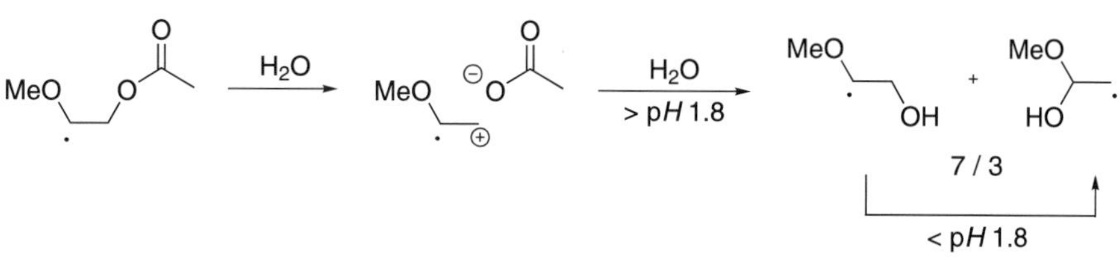

Scheme 3 Chemistry of the β-acetoxy-α-methoxyethyl radical

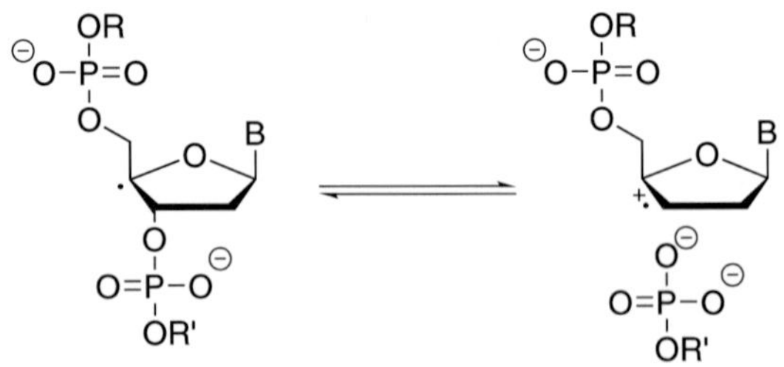

Scheme 4 Expulsion of a phosphate group in the cleavage of an oligonucleotide

Scheme 5 Rearrangement of β-(acyloxy)alkyl radicals

At one time considered as two distinct reactions occurring by different mechanisms [51], the fragmentations of Scheme 2 and the rearrangments of Scheme 5 are now seen as different facets of the same fundamental heterolysis of β-substituted alkyl radicals into alkene radical cations, with the eventual outcome determined by the reaction conditions [52].

3
Structure

The structure of alkene radical cations, planar or twisted, has been controversial. However, on the basis of a great number of sometimes conflicting experimental and theoretical studies, it is generally accepted that the parent ethylene radical cation is significantly twisted so as to permit hyperconjugative stabilization (Fig. 1). As the degree of substitution increases, enabling hyperconjugative stabilization from the substituents, the degree of twisting is reduced. Thus, the ethylene radical cation is considered to be twisted by approximately 25°, whereas the trimethylethylene and tetramethylethylene analogs are essentially planar [53–57]. An X-ray crystal structure of the sesquihomoadamantane radical cation (**1**) showed a twist of 29° over that in the essentially planar alkene precursor [58]. Careful analysis of the crystal structure provided evidence for hyperconjugative stabilization by the β-C – C bonds in the twisted alkene radical cation [58]. Nelsen, Williams, and coworkers showed the bicyclo[2.2.2]oct-2-ene radical cation (**2**) to be significantly more twisted than the more highly substituted 2,3-dimethyl analog (**3**), which can achieve hyperconjugative stabilization in its planar form due to the presence of the methyl groups [59]. ESR studies by Gerson and cowor-

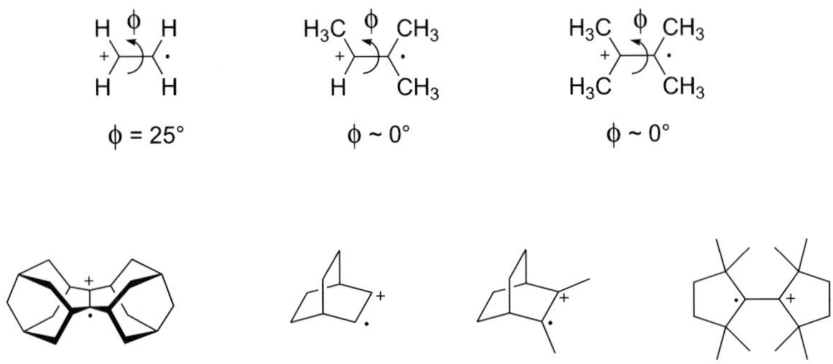

1, $\phi = 29°$ **2**, $\phi = 11–12°$ **3**, $\phi = 5.7°$ **4**

Fig. 1 Twisting in alkene radical cations

kers revealed a series of highly sterically hindered bicycloalkylidene radical cations (**4**) to be twisted, an observation which was attributed to the relief of steric strain [60].

Nelsen and coworkers determined a barrier to inversion through the planar form in **2** and **3** to be approximately 2 kcal mol^{-1} by variable temperature ESR spectroscopy [59]. Gerson and coworkers found, also by ESR spectroscopy, that the frequency of electron exchange between the two sites in **4**, which is equivalent to rotation about the central bond, can vary between $< 10^6$ and $> 10^9$ s^{-1} depending the degree of steric hindrance to planarity [60]. Recent calculations also provide very small barriers to inversion through the planar form [56, 57]. It is apparent, therefore, that for most synthetic purposes most alkene radical cations can be considered as essentially planar with effective delocalization over the two sp^2-hybridized C atoms, and they will be considered as such in this chapter.

4 Mechanistic Underpinnings and Kinetic Data

The first direct observation of an alkene radical cation arising from heterolysis of a β-substituted alkyl radical was made by the Schulte-Frohlinde group, who recorded the ESR spectrum of the 1,1-dimethoxyethene radical cation on generation of the 1,1-dimethoxy-2-acetoxyethyl radical under pulse radiolytic conditions [61]. Using the technique of pulse radiolytic radical generation and time-resolved conductimetry, the German group amassed a large amount of kinetic data on the fragmentation of β-substituted alkyl radicals in aqueous solution [62, 63], some of which are collected in Table 1, with more to be found in previous reviews [51]. Further kinetic data on the fragmentation of DNA-like C4′ radicals were acquired by the Giese group using classical competition radical kinetics [64–67]. Substituent effects on the fragmenta-

Table 1 Rate constants for the fragmentation of β-substituted alkyl radicals

Precursor radical	Solvent	pH	Method[a]	k (s^{-1})	Refs.
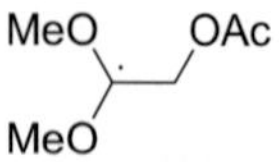	H_2O	slightly acidic	A	$\geq 10^6$	[61]
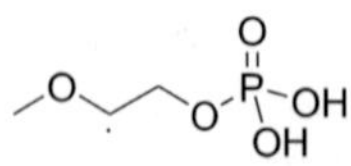	H_2O		A	$\sim 10^6$	[62]
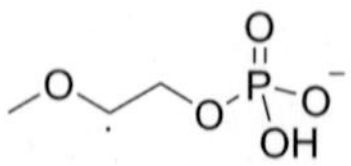	H_2O		A	$\sim 10^3$	[62]
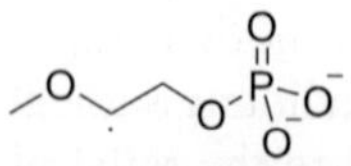	H_2O		A	< 1	[62]
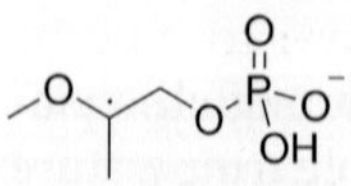	H_2O		A	$\sim 10^6$	[62]
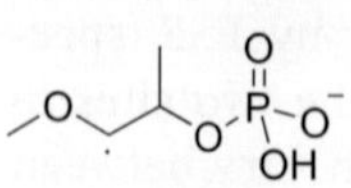	H_2O		A	$\sim 10^6$	[62]
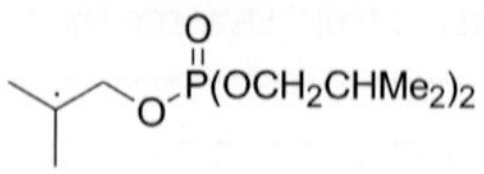	H_2O	4.5 – 5	A	1.4×10^4	[63]
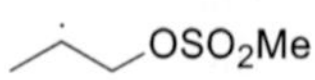	H_2O	4.5 – 5	A	2.0×10^5	[63]
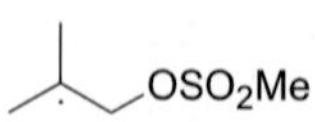	H_2O	4.5 – 5	A	$\geq 10^6$	[63]
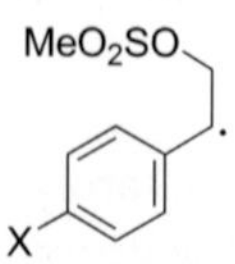					
X = OMe	TFE[b]		B	6.7×10^7	[68]
X = Me	TFE[b]		B	4.6×10^6	
X = Me	HFIP[b]		B	5.7×10^7	
X = H	HFIP[b]		B	5.0×10^5	

[a] Method A: radical generation by pulse radiolysis in conjunction with time-resolved conductivity; Method B: radical generation by laser flash photolysis in conjunction with time-resolved absorption spectroscopy. All the kinetic experiments were run between 20 and 25 °C.

[b] TFE: 2,2,2-trifluoroethanol, HFIP: 1,1,1,3,3,3-hexafluoro-2-propanol

tion of 2-(mesyloxy)-1-phenylethyl were studied by Cozens and coworkers in a range of polar organic solvents [68]. Overall, these kinetic measurements show the expected trends for a heterolytic fragmentation, with better leaving groups departing more quickly and with fragmentation assisted by the presence of electron-donating groups on the nascent alkene radical cation. The influence of more remote groups, such as the base in nucleotide C4′ radical fragmentation, on the rate of fragmentation has also been studied [69].

Further evidence for the formation of alkene radical cations derives from the work of Giese, Rist, and coworkers who observed a chemically induced dynamic nuclear polarization (CIDNP) effect on the dihydrofuran **6** arising from fragmentation of radical **5** and electron transfer from the benzoyl radical within the solvent cage (Scheme 6) [67].

Much kinetic data have also been compiled for the β-(acyloxy)alkyl, β-(phosphatoxy)alkyl, and related radical rearrangements by both competition kinetic methods and kinetic ESR, a selection of which is given in Table 2 with more to be found in a previous review [51]. Classical physical organic structure–reactivity relationships revealed both the acyloxy and the phosphatoxy rearrangements to be accelerated by the presence of electron-withdrawing groups on the migrating ester, and by electron-donating groups on the carbon skeleton [70–72]. The acyloxy migration of salicylate esters is significantly accelerated in the presence of Lewis acids, indicative of stabilization of the migrating carboxylate through chelate formation [73].

Newcomb, Crich, and coworkers studied the acyloxy and phosphatoxy alkyl rearrangements in a range of solvents by means of time-resolved laser flash photolysis, with UV detection of the rearranged benzylic radicals in nonpolar solvents [74]. In polar solvents, on the other hand, these workers noted and quantified the appearance of styrene radical cations arising from the heterolytic cleavage reaction. A plot of the log of the rate constant for either rearrangement to the benzylic radical, or fragmentation to the styrene radical cation, against the E_T30 solvent polarity scale [75] was linear [76–79]. Combined with the closely related entropy terms (log *A*)

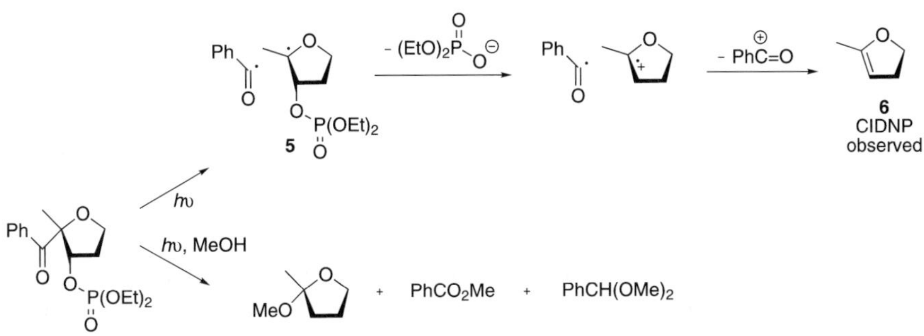

Scheme 6 Observation of a CIDNP effect on fragmentation of radical **5**

Table 2 Rate constants for the rearrangements of β-substituted alkyl radicals

Precursor radical	Product radical	Solvent	T (°C)	k (s^{-1})	Method[a]	Refs.
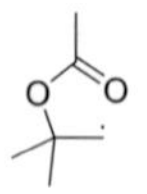	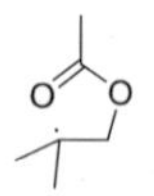	C_6H_6	75	6.2×10^3	A	[72]
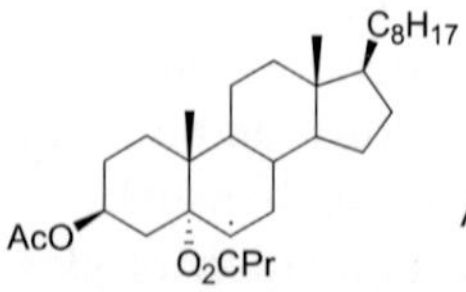	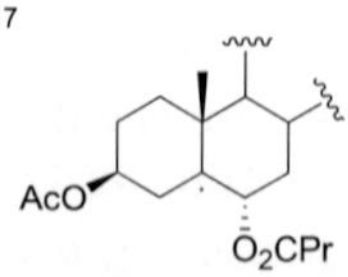	C_6H_6	75	1.9×10^6	A	[121]
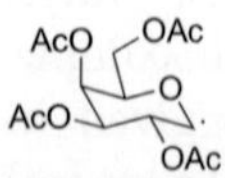	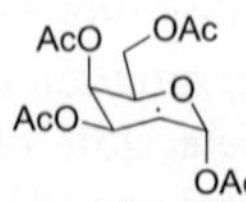	C_6H_6	75	5.4×10^3	B	[118]
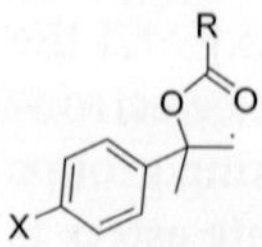	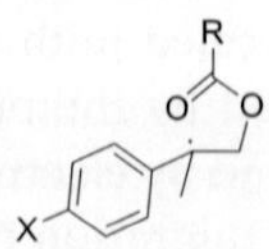					
X = H, R = n-C_3H_7		C_6H_6	75	6.2×10^4		
X = MeO, R = n-C_3H_7		C_6H_6	75	1.7×10^5	C	[71]
X = CN, R = CF_3		C_6H_6	75	2.5×10^6		
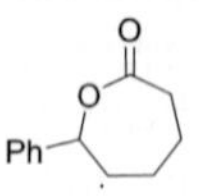	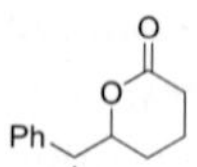	C_6H_6	80	1.7×10^6	C	[125]
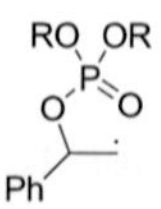	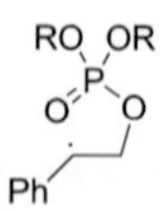					
R = Ph		C_6H_6	80	8.0×10^5		
R = 4-MeOC_6H_4		C_6H_6	80	5.3×10^5	C	[70]
R = 4-$CF_3C_6H_4$		C_6H_6	80	1.2×10^7		
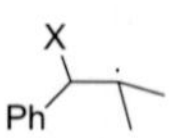	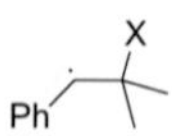					
X = $(PhO)_2P(O)O$		C_6H_6	20	1.2×10^6		
X = $(PhO)_2P(O)O$		MeCN	20	1.8×10^7	D	[74]
X = $(EtO)_2P(O)O$		MeCN	20	$6–7 \times 10^4$		
X = CF_3CO_2		MeCN	20	6.2×10^6		

[a] Method A: radical clock reaction (Bu_3SnH, AIBN); Method B: radical generation by Bu_3SnH/AIBN in conjunction with electron spin resonance; Method C: radical clock reaction (Bu_3SnH, PhSeSePh, AIBN); Method D: radical generation by laser flash photolysis in conjunction with time-resolved absorption spectroscopy

for the fragmentation and rearrangement processes, this led to the conclusion that the rearrangement and fragmentation reactions proceed via a common rate-determining step, namely heterolysis to give a contact alkene radical cation/anion pair. In nonpolar solvents this contact alkene radical cation/anion pair immediately collapses to the observed rearranged radical, whereas in polar solvents the radical cation is sufficiently long-lived for direct observation. This unified mechanism (Scheme 7), a version of which was first advanced by Sprecher [80] and which is nothing more than the open-shell equivalent of the classical ion-pair mechanism for solvolysis first advanced by Winstein [81–83], provides the basis for the studies described in this chapter.

Subsequent work by the Newcomb group, using a combination of classical competition kinetics with trapping by thiophenol and ultrafast radical reporter groups, has enabled rates for some heterolysis reactions to be determined in nonpolar organic solvents (Table 3) [84–87]. The apparent discrepancies between the rate constants reported in Tables 1 and 3 are suggested to arise from the kinetic method employed: the results presented in Table 3 relate directly to the alkene radical cation, whereas those in Table 1 are indirect and arise from an increase in conductivity of the solvent system. It is possible that this increase in conductivity does not occur until trapping of the alkene radical cation by water, followed by deprotonation, which means that the values reported in Table 1 are composite rate constants containing the rates of fragmentation, trapping, and deprotonation [84].

For the 2-methyl-3-phenyl-3-(diphenylphosphatoxy)-2-propyl radical rate constants were obtained for the complete set of processes, including fragmentation to the contact ion pair, collapse of the contact ion pair to the rearranged

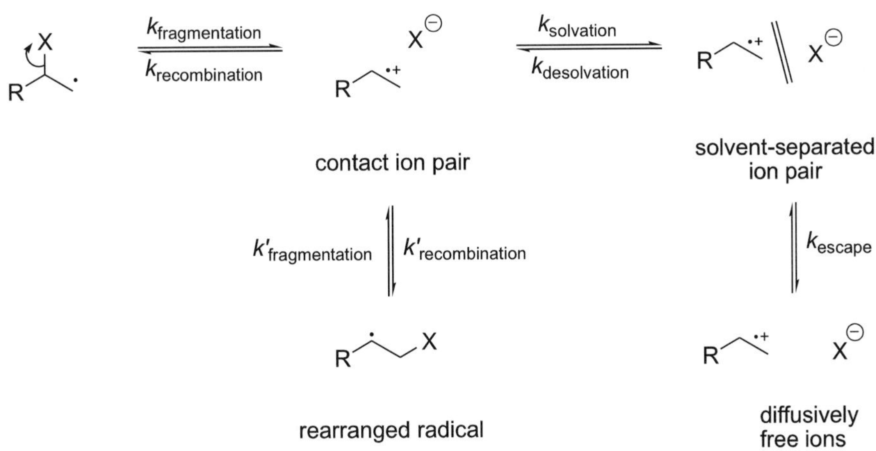

Scheme 7 Unified mechanism of rearrangement and fragmentation of β-substituted radical

Table 3 Rate constants for the fragmentation of β-substituted alkyl radicals in organic solvents

Precursor radical	Solvent	T (°C)	k (s^{-1})	Method [a]	Refs.
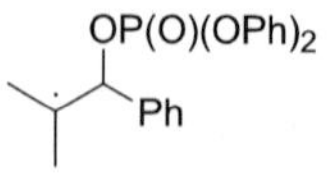	TFE[b] (5%) in toluene	20	9×10^7	A	[78]
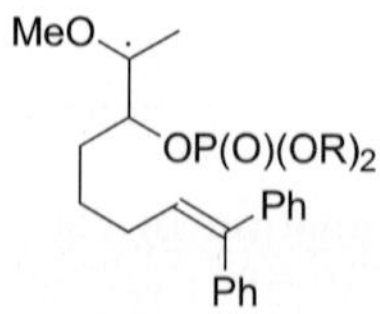					
R = Et	CH_3CN	20	3.9×10^7	A	[86]
R = Ph	THF	20	$> 2.0 \times 10^8$		
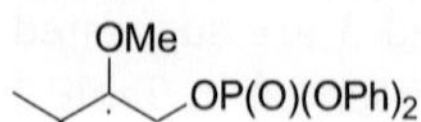	CH_3CN	23	8×10^6	A	[87]
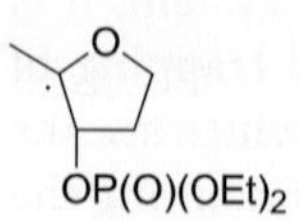	MeOH	25	$> 3.0 \times 10^9$	B	[67]
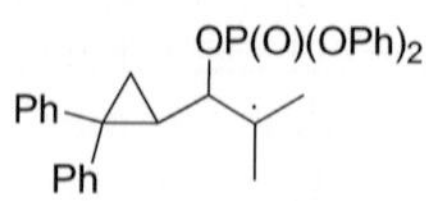	toluene CH_3CN	22 22	7.6×10^6 1.4×10^8	C	[84]
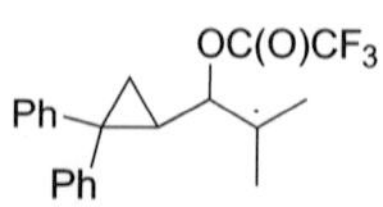	toluene CH_3CN	22 22	1.5×10^6 4.3×10^7	C	[84]

[a] Method A: radical generation by laser flash photolysis in conjunction with time-resolved absorption spectroscopy; Method B: time-resolved CIDNP and competitive kinetic experiments; Method C: radical generation by laser flash photolysis in conjunction with competitive kinetic experiments (trapping by thiophenol and ultrafast radical reporter groups)
[b] TFE: 2,2,2-trifluoroethanol

radical, and solvation of the contact ion pair to the solvent-separated ion pair in a range of solvents (Scheme 8), from which ion pair lifetimes could be estimated [78]. In general the ion pair lifetimes and rates of equilibration with solvent agree with those found previously for radical cation/radical anion pairs formed by photostimulated electron transfer [88]. The very rapid collapse of the ion pairs to starting radicals and rearranged radicals, compared to the rates of rearrangement observed in nonpolar solvents (Table 2), indi-

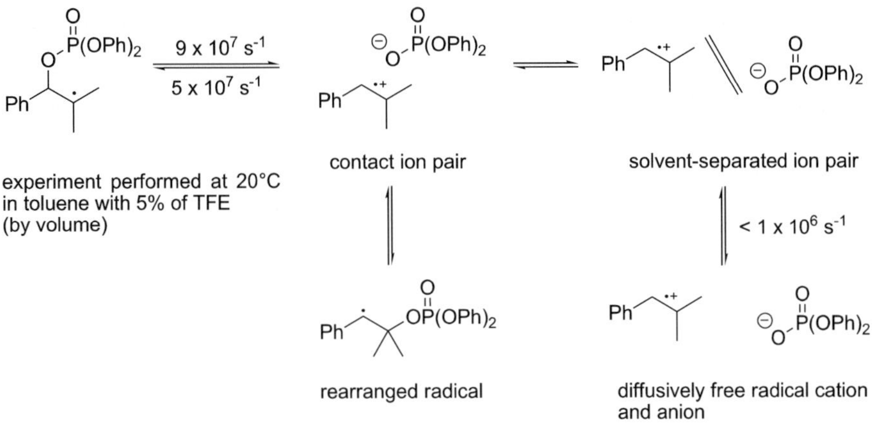

Scheme 8 Fragmentation, rearrangement, and solvation processes of 2-methyl-3-phenyl-3-(diphenylphosphatoxy)-2-propyl radical and associated contact ion pair

cates that the rearrangement can be reliably taken to represent the rates of fragmentation to the contact ion pair.

Relatively few kinetic data are available for the carbon–carbon bond forming reactions of alkene radical cations. Nevertheless, rate constants for the cyclization illustrated in Scheme 9, with generation of the alkene radical cation by the fragmentation method, have been measured. These cyclization rate constants are significantly faster than those of the corresponding neutral radicals [89].

It is important to note in planning synthetic schemes that alkene radical cations are extremely acidic substances. In the context of their generation

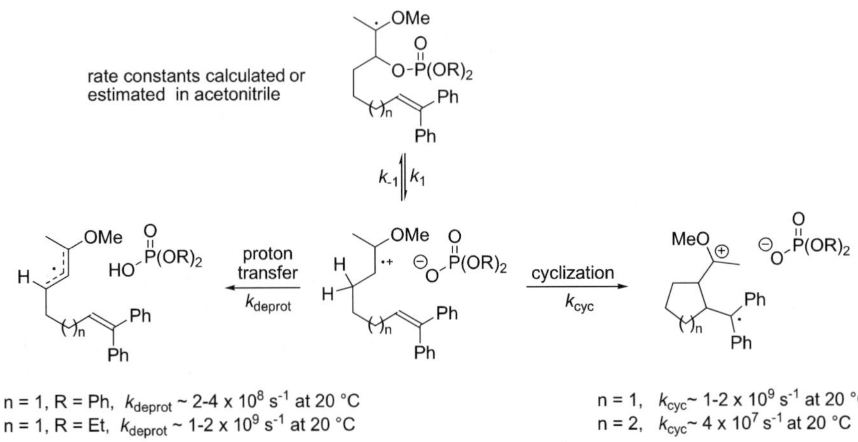

Scheme 9 Cyclization and deprotonation of an alkene radical cation

by the fragmentation of β-substituted alkyl radicals, they may be deprotonated in the contact ion pair by the counterion to give allyl radicals [86, 90]. For example, the radical cation of Scheme 9 is deprotonated by the diphenyl phosphate anion with rate constants approaching those for cyclization. With the more basic diethyl phosphate anion, deprotonation is even faster and is comparable to cyclization [86]. Notably, it has been found that tetrahydrofuran may serve as a base for the deprotonation of alkene radical cations, with a pseudo-first-order rate constant of $1.2 \times 10^7\ s^{-1}$ for the β-methoxy-β-methylstyrene radical cation, when used as solvent for the generation of these species [85].

Although cycloaddition reactions have yet to be observed for alkene radical cations generated by the fragmentation method, there is a very substantial literature covering this aspect of alkene radical cation chemistry when obtained by one-electron oxidation of alkenes [2–16, 18–26, 28–31]. Rate constants have been measured for cycloadditions of alkene and diene radical cations, generated oxidatively, in both the intra- and intermolecular modes and some examples are given in Table 4 [91, 92].

There are extensive kinetic data on the rates of trapping of alkene radical cations by external nucleophiles (Table 5), with the variation between research groups most probably attributable to the kinetic method employed. Schulte-Frohlinde and coworkers determined rate constants for the addition of hydroxide and hydrogen phosphate to the 1,1-dimethoxyethene radical cation by time-resolved conductimetry [93]. Johnston and coworkers measured rate constants for the addition of a variety of anionic and neutral nucleophiles to substituted styrene radical cations, generated by photooxidation, using time-resolved laser flash photolysis with UV detection [92, 94], as compiled in several reviews [95, 96]. More recently, Newcomb and coworkers, employing alkene radical cations generated by the fragmentation method under laser flash photolytic conditions, determined rate constants for the addition of acetonitrile, methanol, and water to various alkene radical cations, and drew attention to the reversibility of the alcohol addition [84, 86].

The regiochemistry of nucleophilic addition to alkene radical cations is a function of the nucleophile and of the reaction conditions. Thus, water adds to the methoxyethene radical cation predominantly at the unsubstituted carbon (Scheme 3) to give the β-hydroxy-α-methoxyethyl radical. This kinetic adduct is rearranged to the thermodynamic regioisomer under conditions of reversible addition [33]. The addition of alcohols, like that of water, is complicated by the reversible nature of the addition, unless the product distonic radical cation is rapidly deprotonated. This feature of the addition of protic nucleophiles has been studied and discussed by Arnold [5] and Newcomb [84, 86] and their coworkers.

Using alkene radical cations generated under photostimulated electron-transfer conditions, Arnold and coworkers showed that the addition of an-

Table 4 Rate constants for cycloadditions with alkene radical cations [a]

Starting radical cation	Trap	Product	k [b]	Refs.
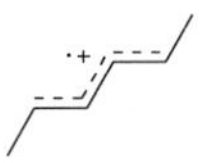	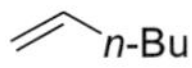	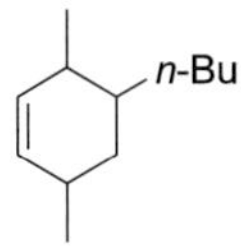	$< 3.0 \times 10^6$	[92]
	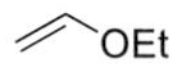	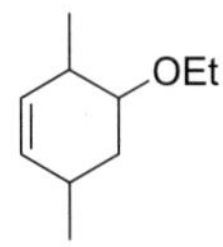	3.2×10^9	[92]
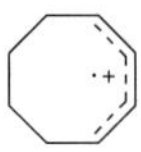		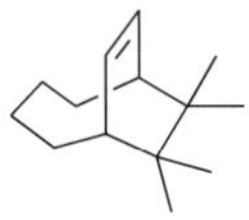	5.1×10^9	[92]
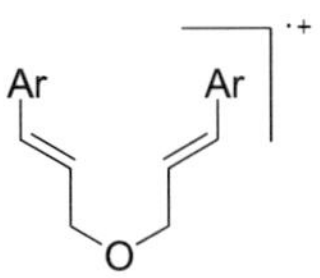 Ar = 4-methoxyphenyl		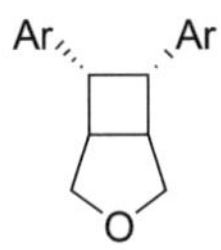	1.2×10^9	[91]
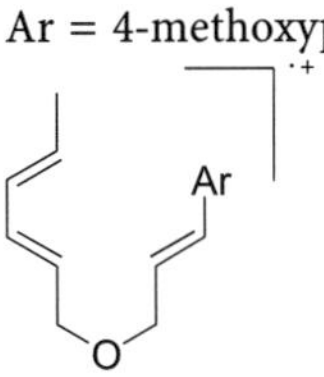 Ar = 4-methoxyphenyl		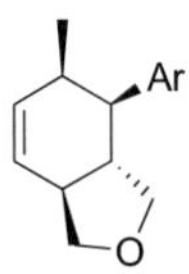	3.0×10^8	[91]

[a] Radical generation by laser flash photolysis in conjunction with time-resolved absorption spectroscopy. The experiments were run in acetonitrile at room temperature.
[b] $M^{-1}\,s^{-1}$ for bimolecular reactions and s^{-1} for unimolecular reactions

ionic nucleophiles, such as cyanide and fluoride, is under kinetic control and that the product ratio is determined by steric and polar factors rather than by the relative stabilities of the radicals formed [5]. The attack of hydroxide and hydrogen phosphate anions on the 1,1-dialkoxyethene radical cations was studied by Schulte-Frohlinde and coworkers, with ESR detection of the resulting radicals, although no clear guidelines were given for regioselectivity [93]. Acetonitrile appears to function similarly; the distonic radical nitrilium ion is subject to a range of subsequent reactions [5]. Overall, the picture that emerges for kinetically controlled additions is one of addition to the least substituted terminus of simple alkene radical cations.

Table 5 Rate constants for nucleophilic addition to alkene radical cations

Starting radical cation	Nucleophile	Solvent	k (M^{-1} s^{-1})	Refs.
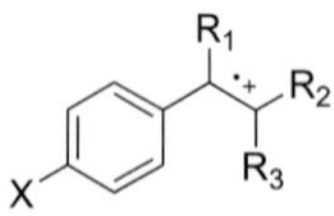				
X = H, $R_1 = R_2 = R_3 = H$		TFE [d]	1.8×10^8	[94] [a]
X = Me, $R_1 = R_2 = R_3 = H$		TFE [d]	5.9×10^6	[94] [a]
X = OMe, $R_1 = R_2 = R_3 = H$	MeOH	$CH_3CN/H_2O(1/4)$	3.0×10^4	[94] [a]
X = H, R_1 = Me, $R_2 = R_3 = H$		TFE [d]	1.9×10^8	[94] [a]
X = H, R_1 = H, R_2 = Me, R_3 = H		TFE [d]	9.7×10^6	[94] [a]
X = H, R_1 = H, $R_2 = R_3$ = Me		TFE [d]	2.0×10^5	[94] [a]
	MeOH	CH_3CN	1.8×10^8	[92] [a]
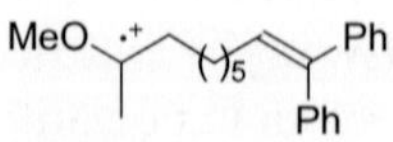	MeOH	CH_3CN/TFE [d] 2.5%	1.4×10^6	[86] [b]
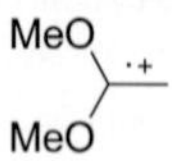	OH^-	H_2O	4.2×10^9	[93] [c]
	HPO_4^{2-}	H_2O	9.0×10^5	[93] [c]
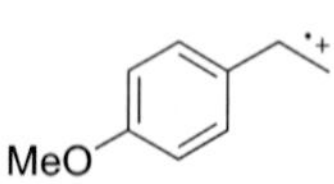	N_3^-	CH_3CN	4.2×10^{10}	[94] [a]
	Br^-	CH_3CN	4.0×10^{10}	[94] [a]
	n-$BuNH_2$	CH_3CN	2.5×10^9	[96] [a]
	N_3^-	CH_3CN/H_2O (1/4)	1.2×10^{10}	[94] [a]
	Br^-	CH_3CN/H_2O (1/4)	2.2×10^6	[94] [a]
	MeOH	CH_3CN/H_2O (1/4)	3.0×10^4	[94] [a]
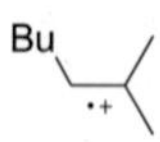	CH_3CN	CH_3CN	1.0×10^6	[84] [b]

[a] Radical generation by photoionization or photosensitization, using time-resolved laser flash photolysis with UV detection
[b] Radical generation by laser flash photolysis in conjunction with time-resolved absorption spectroscopy
[c] Radical generation by pulse radiolysis in conjunction with time-resolved conductivity
[d] TFE: 2,2,2-trifluoroethanol

5 Computational Studies

The mechanisms of the β-(acyloxy)alkyl, and later the β-(phosphatoxy)alkyl rearrangements have been the subject of considerable effort by computational

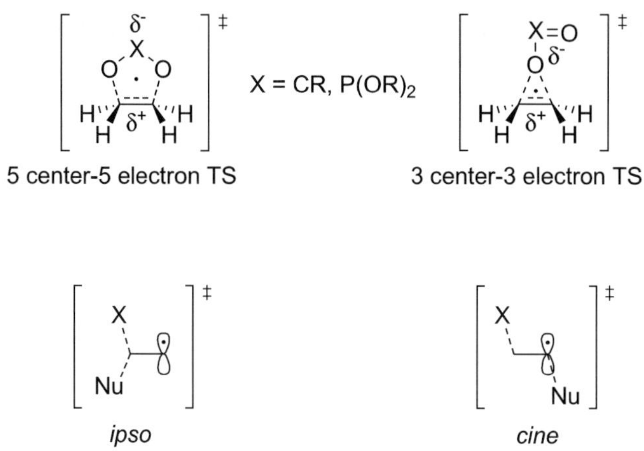

Fig. 2 Computed concerted transition states for rearrangement and substitution reactions

chemists, beginning with the early work of Radom [97], and continuing with the extensive studies of Zipse [98]. At the time of the last review of the area in 1997 [51] the computational work largely supported, indeed was an important factor in, the then prevailing view of two concerted pathways. Thus, the generally slower acyloxy shift, with its high predilection for inversion of the carboxyl oxygens, was predicted to take place through a five-center–five-electron cyclic transition state with significant charge separation (Fig. 2). This cyclic transition state is distinct from the possibility of a 1,3-dioxolan-2-yl radical intermediate which had been eliminated conclusively by experiment [99]. The more rapid phosphatoxy shift, on the other hand, was computed to involve a three-center–three-electron cyclic transition state (Fig. 2), with a somewhat greater separation of charge, as the main pathway with a minor component of the slower five-center–five-electron transition state, in agreement with the observed preponderance of a 1,2-shift. However, as the calculations have evolved the degree of charge separation has increased to the extent that the most recently computed 75% charge separation in the phosphatoxy shift is tantamount to a contact ion pair [98, 100], even if this is not yet the case for the acyloxy migration [98, 101]. Pathways have also been computed for the concerted displacement of leaving groups in both the *ipso* and *cine* modes (Fig. 2) [102, 103] but, for those cases which have been tested experimentally [66, 102, 104], the evidence favors a stepwise mechanism via a contact alkene radical cation/anion pair.

6
Suitable Radical Precursors

In designing preparative radical ionic chain reactions, including the fragmentation approach to alkene radical cations, careful choice of the radical precursor is required. This is especially the case when the reaction sequence envisaged includes an intramolecular nucleophilic attack on the alkene radical cation. In such cases the radical precursor to the alkene radical cation must be such that it is not susceptible to premature reaction with the nucleophile. This effectively excludes the use of the standard alkyl halide/stannane chain sequences in all but the simplest systems. Alkyl phenyl selenides are convenient radical precursors in conjunction with stannanes [105], having comparable reactivity to the corresponding bromides [106]. Unfortunately, 2-phenylselenoalkyl phosphates and mesylates are unstable with respect to elimination to the alkene via the intermediacy of episelenonium ions (Crich et al., unpublished results) [107]. This decomposition pathway, which also holds for the corresponding sulfides, prevents the use of selenides in this chemistry unless the system is constrained so as to prevent episelenonium ion formation. The use of *O*-acyl thiohydroxamates, or Barton or pyridinethioneoxycarbonyl (PTOC) esters [108–110], has found wide application in the kinetic work of Newcomb in this area, but has limited range in preparative sequences owing to the highly activated carbonyl group, which renders it incompatible with many nucleophiles. Intramolecular hydrogen abstraction has proven to be a useful tool with appropriately designed systems [111]. Another useful tool, applied in the earliest work of the Giese group on model DNA C4′ radicals, is the addition of thiyl radicals to the terminus of allylic phosphates [65]. However, this protocol suffers from the reversibility of the thiyl radical addition to the alkene, with the result that the fragmentation reaction is influenced by the type and concentration of thiol as well as by the initiator system [112]. The most successful precursor to date has been the tertiary nitro group. This group is moderately reactive toward tin hydrides [113, 114], and takes advantage of the facile assembly of β-nitroalcohols and their esters by means of the Henry reaction. The nitro group has the additional advantage of being powerfully electron-withdrawing, which helps to stabilize β-nitrophosphates and related substrates against premature solvolysis before the radical chemistry can be undertaken. One disadvantage of the nitro group as radical precursor is the ease of elimination of the β-phosphate or other leaving group with the consequence that, for all practical purposes, the system must be fully substituted so as to prevent formation of nitronate anions. A similar restriction relating to elimination pertains to the use of *O*-acyl thiohydroxamates.

7 Reinterpretation of Ester Rearrangements

The rearrangements of β-(acyloxy), β-(phosphatoxy)alkyl, and related systems have been reviewed [51, 52] and representative kinetic data are given in Table 2 above. As revealed by isotopic labeling experiments, the acyloxy shift proceeds with a very high degree of inversion of the carboxyl oxygens, resulting in a formal 2,3-shift. Related isotopic and stereochemical experiments established the phosphatoxy migration to be predominatly a 1,2-shift. These experiments, along with the suprafacial shift of the migrating group along one face of the carbon framework in cyclic systems and the failure of various crossover experiments, provided the basis for the earlier interpretation of the acyloxy shift as proceeding via a five-center–five-electron concerted pathway. Likewise, the phosphatoxy shift was interpreted in terms of a three-center–three-electron pathway alongside a minor five-electron–five-center pathway [51]. The general mechanism presented in Scheme 7 provides for the reinterpretation of these rearrangements in terms of fragmentation to a contact alkene radical cation/anion, with extremely rapid collapse to the product radical on a timescale faster than equilibration of the ion pair [74]. Thus, in one of the preparatively more significant examples of the acyloxy shift [115–117], in 1-glycosyl radicals a labeled benzoate migrates to the anomeric radical along one face of the pyranose ring with complete inversion of the carboxylate [118], a result which is now best viewed in terms of the mechanism set out in Scheme 10.

The suprafacial shift along the carbon framework is not restricted to cyclic systems but may also prevail in acyclic cases. In the example given in Scheme 11, minimization of dipolar repulsion between the two C – O bonds mandates a preferred conformation of the initial radical, leading to a stereochemically defined alkene radical cation and, ultimately, to a single diastereomer of the product [119].

Examples of the acyloxy shift that proceed with less than 100% inversion of the carboxyl oxygens [120–123] are now best interpreted in terms of looser ion pairs, resulting from more highly stabilized and/or substituted alkene

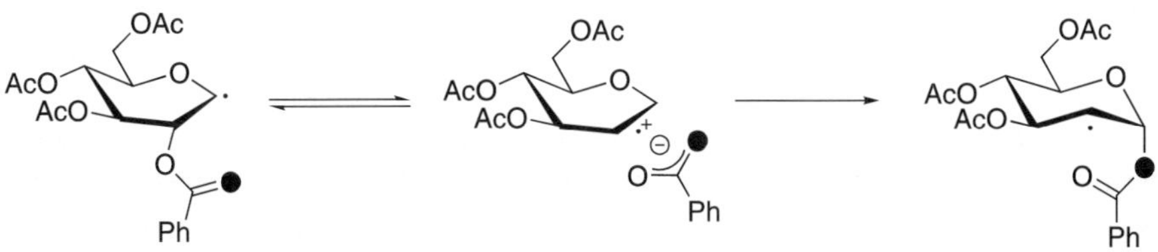

Scheme 10 Migration in 1-glycosyl radicals with a labeled benzoate group

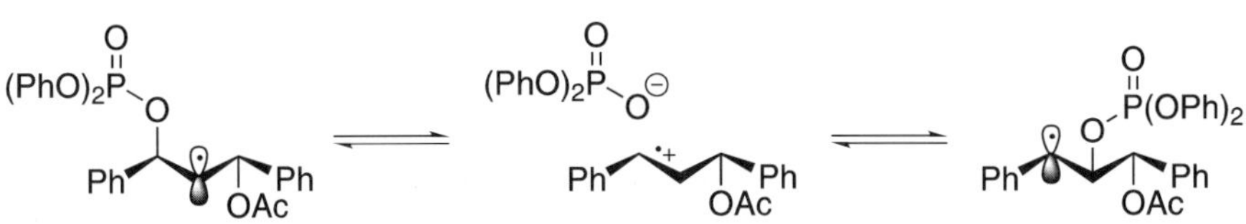

Scheme 11 Suprafacial shift in an acylic system

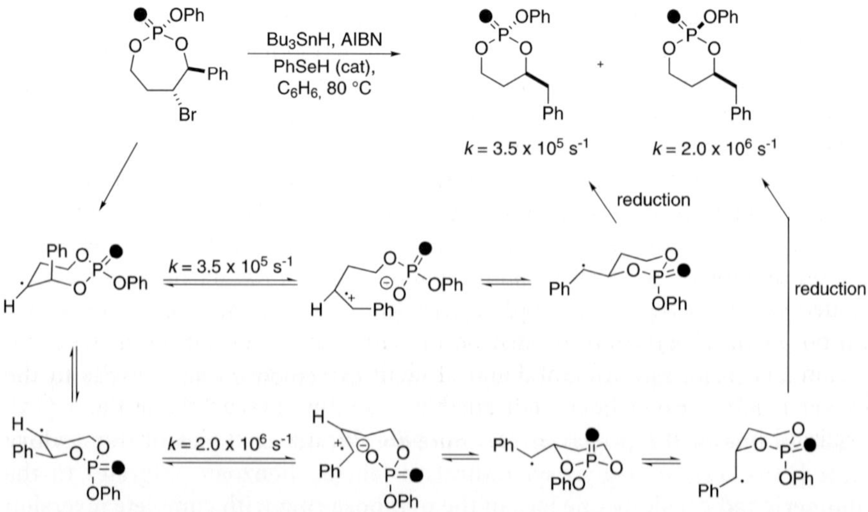

Scheme 12 1,2-phosphatoxy shifts in a cyclic phosphate ester

radical cations. In conformationally constrained systems, such as lactones, the acyloxy shift can be forced into the 1,2-mode of reaction [124, 125], just as the phosphatoxy shift can be compelled to take place via a pure 1,2-shift in the context of cyclic phosphate esters (Scheme 12) [126]. In the context of the generalized mechanism (Scheme 7), these ring contraction experiments again serve to illustrate the high degree of order of the ion pair and its very rapid collapse. In the example given two stereoisomeric products were formed from a single, stereochemically pure substrate, but isotopic labeling experiments revealed complete retention of configuration at phosphorus in both products. This at first sight confusing observation is the result of the fragmentation occurring from two conformations of the initial radical, leading to two contact radical ion pairs differing in the configuration (*E* or *Z*) of the alkene radical cation, both of which collapse instantaneously to the product radical.

Ion pair collapse in the acyloxy migration is so rapid as to preclude nucleophilic trapping of the contact ion pair even by intramolecular nucleophiles, which essentially precludes the use of acetates as leaving groups in tandem rearrangement reactions of the types discussed below [111, 127].

8
Radical Cyclizations

In a rare study of a radical cyclization of fragmentation-derived alkene radical cations, it was discovered that the stereochemistry of the precursor can have significant consequences on the outcome of the reaction. Thus, a *gluco-*

derived substrate underwent a clean, high-yielding transformation to give a single stereoisomeric product (Scheme 13), presumably by the pathway illustrated. In contrast, the *manno* isomer (Scheme 14) required significantly longer reaction times, leading to a complex reaction mixture from which the product was only isolated in very low yield [128].

In the *gluco* case (Scheme 13) the radical cyclization, with its requirement for the formation of a *cis*-fused ring junction [129, 130], takes place uneventfully on the opposite face of the alkene radical cation to the one shielded by the phosphate anion, whereas in the *manno* series cyclization is severely retarded by the presence of the phosphate group above the face of the radical cation on which cyclization must occur. This steric retardation of the cyclization step results in a breakdown of chain propagation and results in the longer reaction times observed. Furthermore, the retardation of the radical cyclization step in the *manno* case enables the alkene radical cation to take

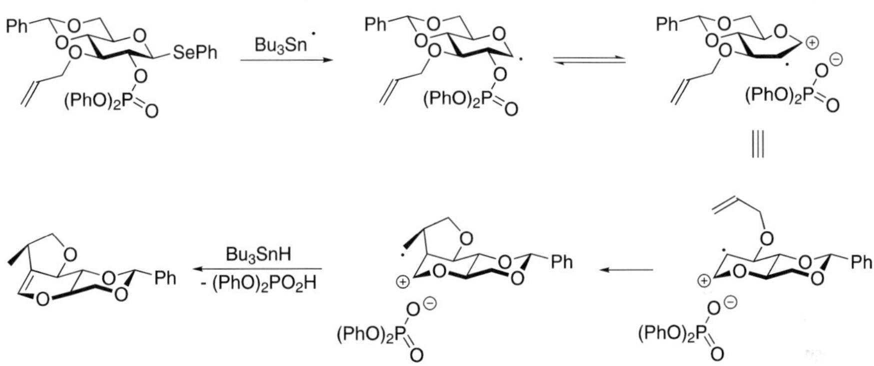

Scheme 13 Radical cyclization of a *gluco*-derived substrate

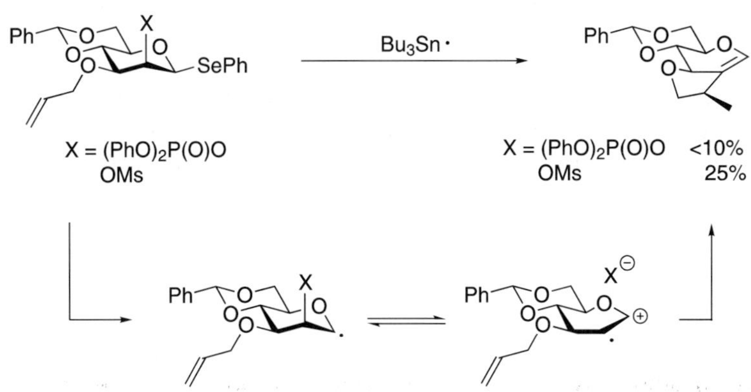

Scheme 14 Radical cyclization of a *manno*-derived substrate

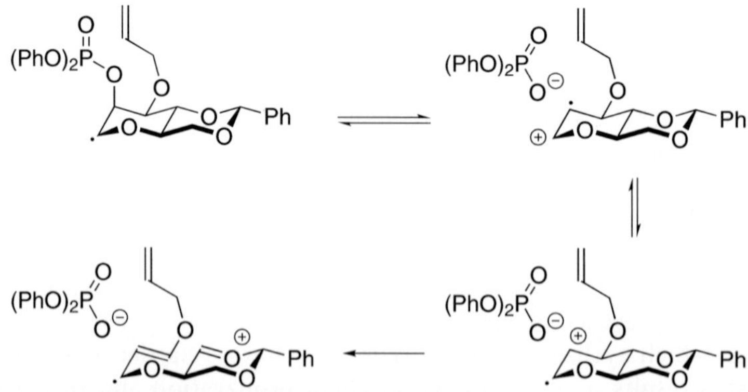

Scheme 15 Fragmentation of an alkene radical cation

part in alternative processes, perhaps including the fragmentation shown in Scheme 15 [128].

Consistent with this argument, replacement of the phosphate in the *manno* series by a mesylate group, with its better leaving group ability and presumably looser ion pair, resulted in a moderately increased yield of cyclization product. Conversely, and again consistent with the mechanism, the exchange of the phosphate for a mesylate in the *gluco* series occasioned no significant change in yield [131]. The degradation of the implied anomeric phosphate to the glycal observed in these reactions is in full accord with earlier studies of 2-*O*-phosphate-substituted anomeric radicals [49, 50, 119, 132].

9 Intermolecular Nucleophilic Trapping

In a rare example of the use of phenylselenides as radical precursors in the generation of alkene radical cations by the fragmentation approach, Giese and coworkers generated a thymidine C3′,C4′ radical cation by expulsion of diethyl phosphate. Trapping experiments were conducted with methanol and with allyl alcohol (Scheme 16), when nucleophilic attack was followed by radical cyclization [66].

The high degree of stereoselectivity observed in the trapping reaction prompted Zipse to propose a double inversion mechanism, taking advantage of his methyleneology principle [103], involving the thymine carbonyl oxygen [102]. However, subsequent work by the Giese group, this time employing the Norrish type I photofragmentation process to generate the initial radical, showed that similarly high facial selectivity is observed in related systems lacking the possibility of the double inversion (Scheme 17) [90, 104].

The allyl alcohol trapping reaction was further studied by Crich and coworkers, who applied the Barton decarboxylation reaction as radical source,

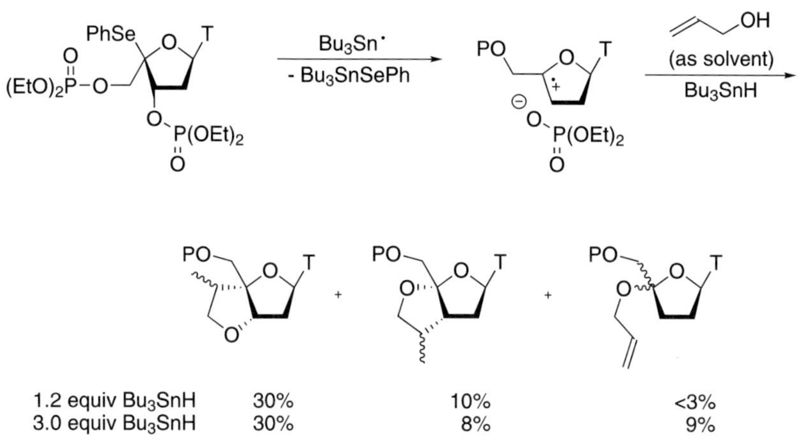

Scheme 16 Generation and trapping of a thymidine C3′,C4′ radical cation

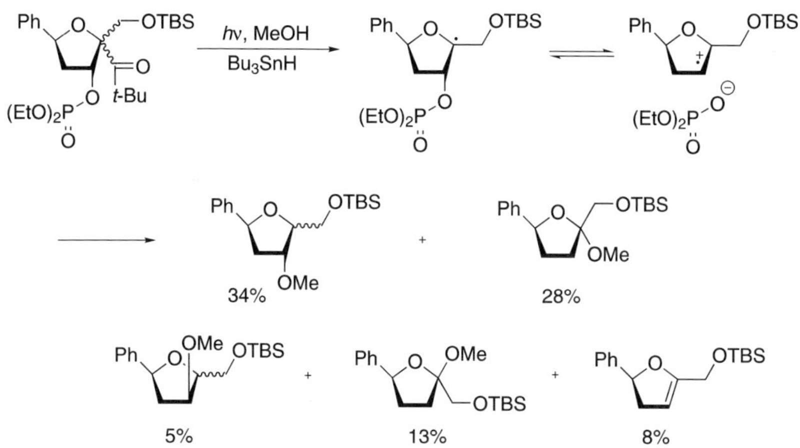

Scheme 17 Radical cation generation and trapping by Norrish type I photofragmentation

and developed the overall process into a preparative method for tetrahydrofuran formation (Scheme 18) [111]. Two regioisomeric precursors to the alkene radical cation were prepared and both led to the same product with comparable yields and stereoselectivity, indicating a common alkene radical cation intermediate [111]. The regioselectivity of the trapping reaction was essentially complete with no isomeric tetrahydrofurans observed, and the stereoselectivity of the radical cyclization step is consistent with other 5-hexenyl cyclizations of benzyl radicals [133].

Crich and Gastaldi investigated the nucleophilic trapping of a dihydronaphthalene radical cation by octyl alcohol and noted that the stereoselectivity of the reaction, while not high, was a function of the substrate stereochemistry (Scheme 19) [134]. In terms of the general mechanism for fragmentation

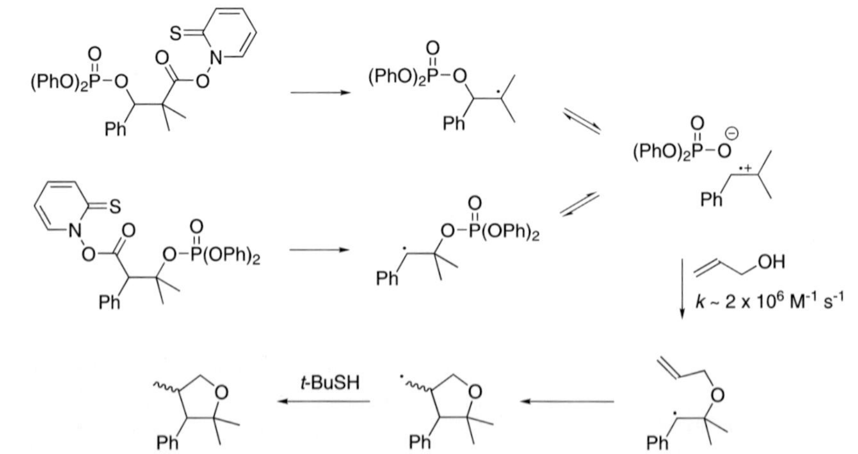

Scheme 18 Tetrahydrofuran formation with the Barton decarboxylation reaction as radical source

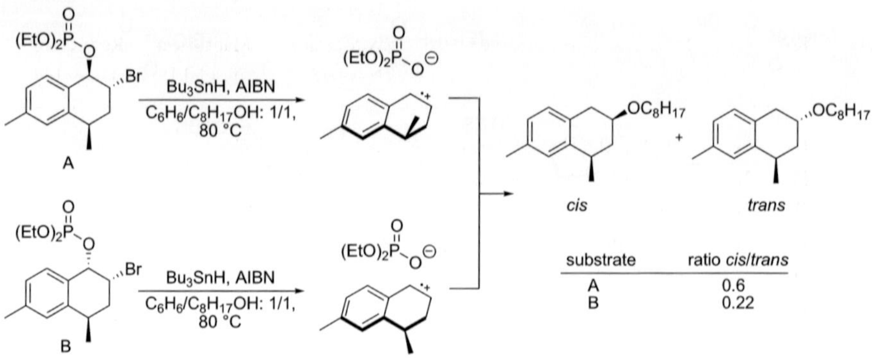

substrate	ratio *cis*/*trans*
A	0.6
B	0.22

Scheme 19 Nucleophilic trapping of a dihydronaphthalene radical cation by octyl alcohol

and substitution (Scheme 7), this chemistry is best interpreted in terms of two diastereomeric alkene radical cation/anion pairs, not unlike the situation with the glucose- and mannose-derived alkene radical cations presented above (Schemes 13 and 14). Further discussion of diastereomeric alkene radical cation/anion pairs is reserved for later in this chapter.

10
Intramolecular Nucleophilic Trapping by Oxygen Nucleophiles

A method for intramolecular nucleophilic attack by alcohols was devised in which the initial radical was generated by a 1,5-hydrogen abstraction

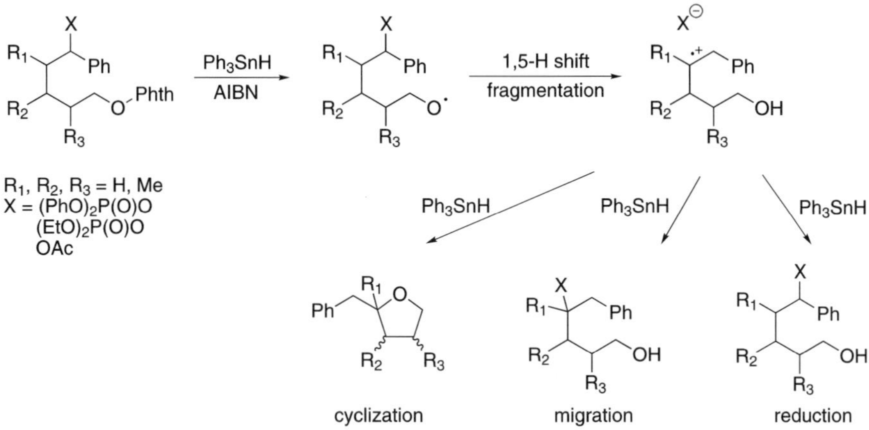

Scheme 20 Radical cation generation by 1,5-hydrogen abstraction and fragmentation

Table 6 Tetrahydrofuran formation

R_1	R_2	R_3	X	Solvent	Cyclization (% yield)	Migration (% yield)	Reduction (% yield)
H	H	H	$(PhO)_2P(O)O$	benzene	95		
H	H	H	$(EtO)_2P(O)O$	benzene	60	25	15
H	H	H	OAc	benzene	0	74 [a]	26 [a]
Me	H	H	$(PhO)_2P(O)O$	benzene	90		
Me	H	H	$(EtO)_2P(O)O$	benzene	85		
Me	H	H	OAc	benzene	0	35 [a]	65 [a]
H	Me	H	$(PhO)_2P(O)O$	benzene	90		
H	H	Me	$(PhO)_2P(O)O$	benzene	92		
Me	H	H	OAc	benzene/ CH_3CN 1/1	0	64 [a]	36 [a]

[a] Products isolated as a mixture of acetates due to scrambling of the acetate group between the two hydroxyls (the scrambling is a post-radical step)

process (Scheme 20, Table 6) [111, 135, 136]: following Kim, the reaction of *N*-alkoxyphthalimides with stannyl radicals served to generate the requisite alkoxy radicals [137]. Interestingly, and in line with precedent [138], the hydrogen-atom abstraction step took place exclusively in the 1,5-manner with no observable encroachment of the 1,6-abstraction even though this would provide a benzylic radical. The influence of the leaving group was examined, with diphenyl phosphate being optimal and acetate being ineffective, because the ring closure step was unable to compete with collapse of the alkene radical

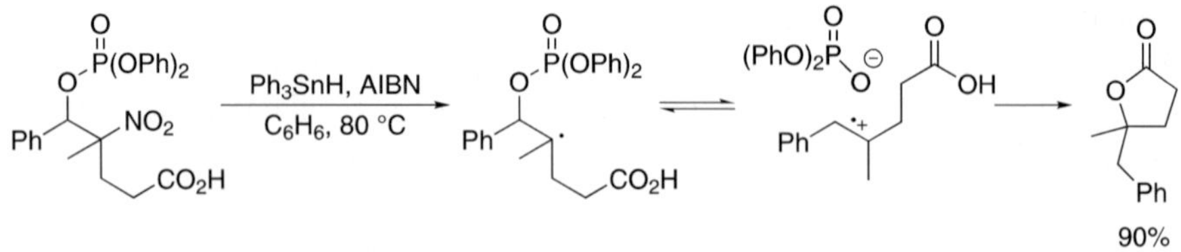

Scheme 21 Nucleophilic cyclization of a carboxylic acid onto an alkene radical cation

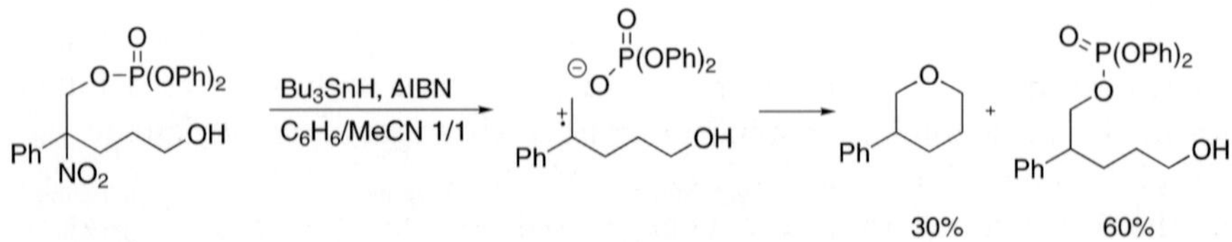

Scheme 22 A 6-*endo* cyclization of an alcohol onto an alkene radical cation/phosphate anion pair

cation/acetate pair to the rearranged radical. As expected, the diethyl phosphates showed intermediate reactivity. Most interestingly, a higher cyclization yield was obtained for the diethyl phosphates when the site of nucleophilic attack was additionally substituted with a methyl group. This phenomenon is due to the greater stability of the more highly substituted alkene radical cation, which retards ion pair collapse to the benefit of the cyclization reaction. Little or no stereoselectivity was observed in these cyclizations.

A γ-lactone was formed in excellent yield by the nucleophilic cyclization of a carboxylic acid onto an alkene radical cation generated from a β-nitrophosphate under tin hydride conditions (Scheme 21) [139]. Related experiments employing the acetate group and an internal carboxylate nucleophile failed, emphasizing the very rapid collapse of the alkene radical cation/acetate ion pair [127].

An example of a 6-*endo* cyclization of an alcohol onto an alkene radical cation/phosphate anion pair has also been described (Scheme 22). In order to bring about fragmentation of the primary alkyl phosphate bond in this reaction it was necessary to work in a 1:1 mixture of benzene and acetonitrile [139, 140].

11 Intramolecular Nucleophilic Trapping by Nitrogen Nucleophiles

The advantage of the nitro group as radical precursor is best seen in the context of intramolecular nucleophilic trapping of alkene radical cations by nitrogen nucleophiles, when no cyclization was observed prior to treatment with tin hydrides and a radical initiator. Substrates may be designed such that the cyclizations take place in either the exo- or the endocyclic mode depending on the substitution pattern of the intermediate alkene radical cation (Schemes 23 and 24) [139–142], as is also the case with oxygen nucleophiles (cf. Schemes 21 and 22).

Some heterocyclic nucleophiles may also be successfully employed in these cyclization reactions (Schemes 25 and 26) [131]. In contrast, no cyclization was observed in an aniline-based system (Scheme 27), which reflects the reduced nucleophilicity of the aniline nitrogen [142].

The real beauty of amines as nucleophiles becomes apparent when substrates are designed so as to incorporate a radical cyclization as a follow up to the nucleophilic trapping process. Cyclization of an allylamine nucleophile onto a conjugated trisubstituted alkene radical cation proceeded in the 5-*exo* mode to give a benzylic radical, which then took part in a 5-*exo*-trig radical ring closure affording a mixture of four stereoisomeric pyrrolizidines (Scheme 28). The four products arise from divergences in the radical cyclization step, with the major product resulting from the expected *trans*-selective cyclization of the benzyl radical through a transition state

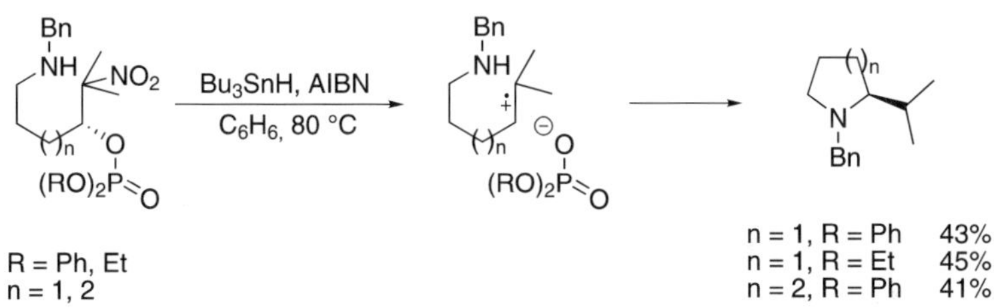

Scheme 23 Cyclization in the exocyclic mode

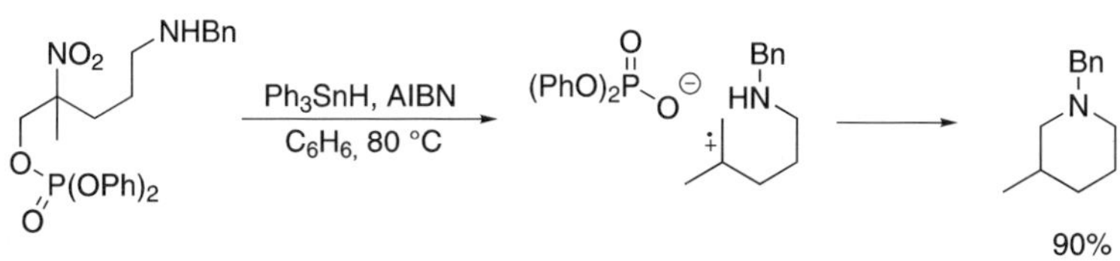

Scheme 24 Cyclization in the endocyclic mode

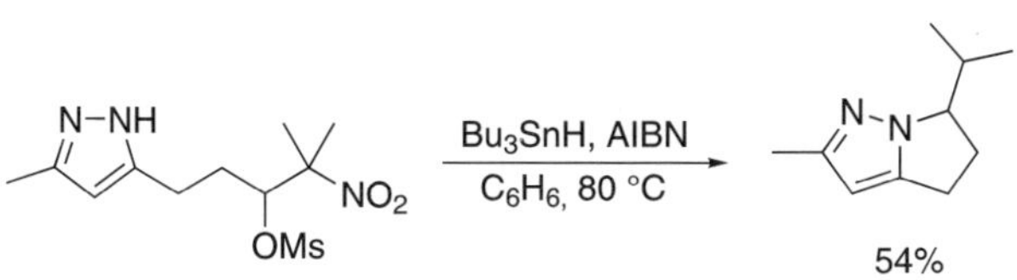

Scheme 25 Pyrazole nitrogen nucleophiles in exocyclic ring closure

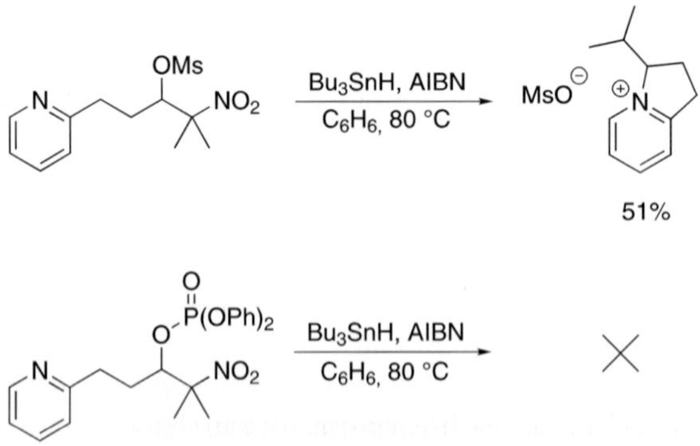

Scheme 26 Pyridine nitrogen nucleophiles in exocyclic ring closure

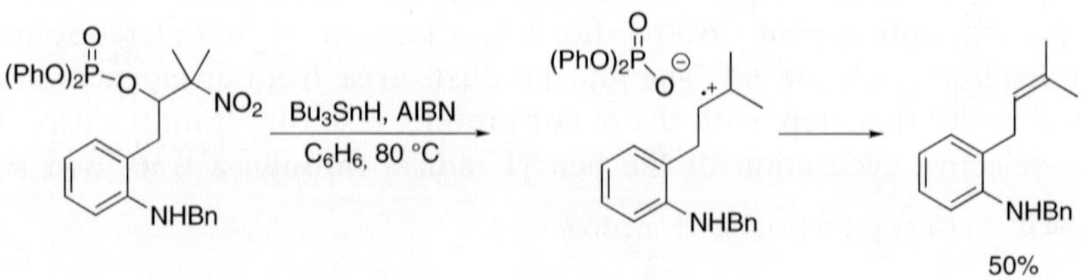

Scheme 27 Absence of cyclization in an aniline-based system

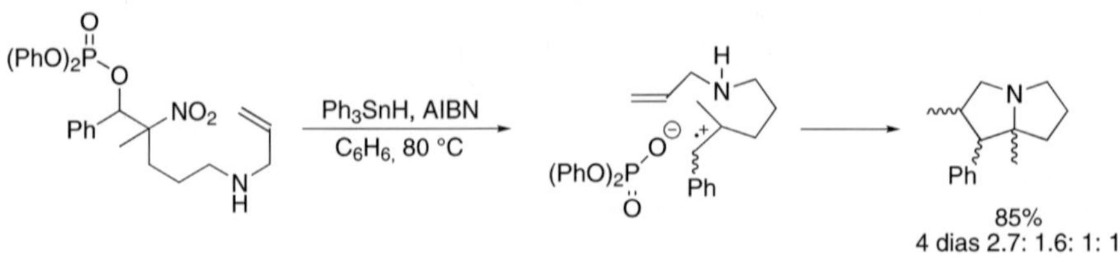

Scheme 28 Pyrrolizidine formation by a tandem cyclization

with the phenyl group on the *exo* face of the incipient bicyclic skeleton (Fig. 3) [139, 143].

Both pyrrolizidines and indolizidines may be similarly formed by cyclization at the less-substituted, internal position of trialkyl-substituted alkene radical cations (Scheme 29) [139, 143]. Related processes featuring *exo*-digonal radical cyclizations have also been described (Scheme 30) [139, 141–143].

An alternative substrate design, in which the alkene radical cation is substituted only at the internal position, forces the nucleophilic cyclization into the endocyclic mode, leading overall to bicyclic systems with a bridgehead nitrogen (Scheme 31) [139, 140].

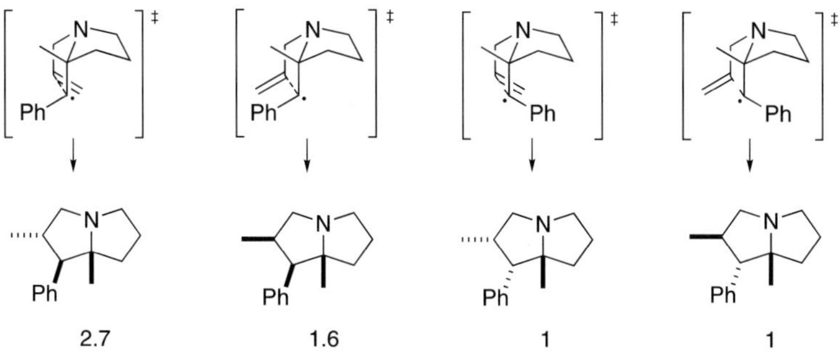

Fig. 3 Transition states leading to diastereomeric pyrrolizidines

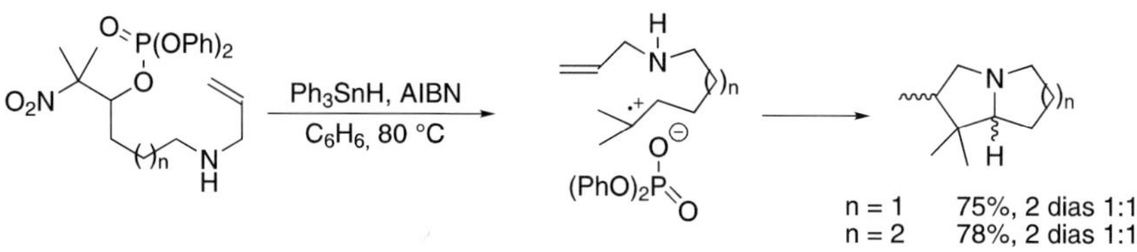

Scheme 29 Further pyrrolizidine formation

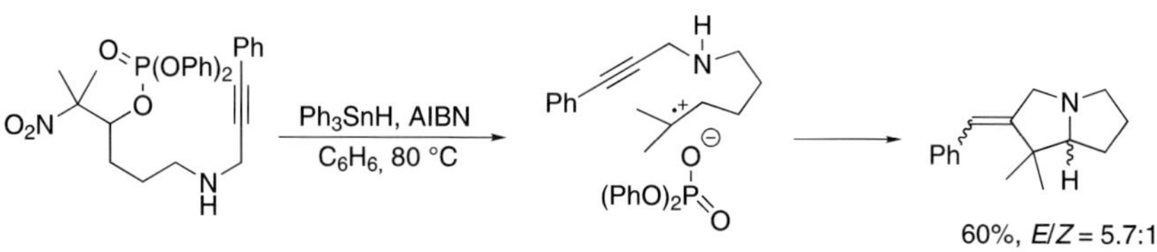

Scheme 30 Nucleophilic trapping followed by *exo*-digonal radical cyclization

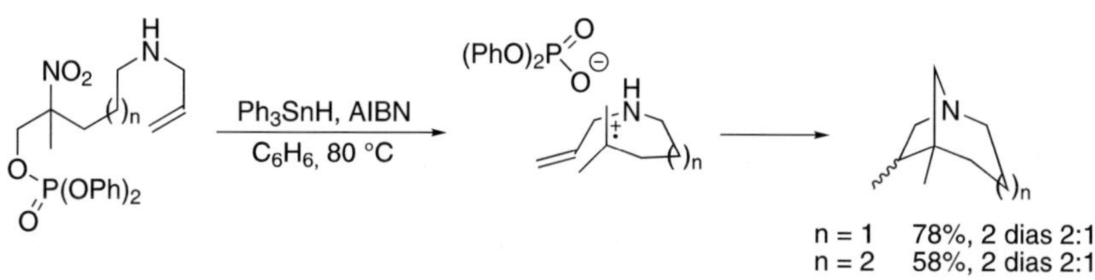

Scheme 31 Nucleophilic cyclization of an alkene radical cation in the endocyclic mode

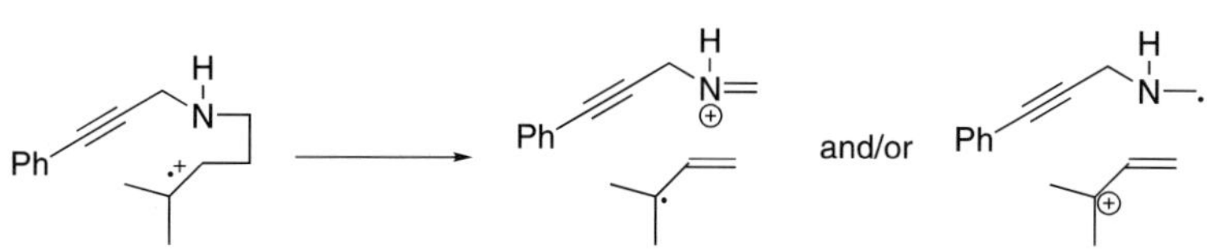

Scheme 32 Alkene radical cation fragmentation

Attempts at 4-*exo* nucleophilic cyclization failed, presumably because of a heterolytic fragmentation of the intermediate radical cation (Scheme 32) [139], not unlike that proposed (Scheme 15) for the decomposition of a mannose-derived alkene radical cation.

12
Diastereoselectivity in Nucleophilic Cyclizations

The ability of nucleophiles to compete with the collapse of the contact alkene radical cation/anion pairs generated by the fragmentation method, along with the various isotopic and stereochemical labeling experiments indicating that the contact ion pairs recombine to give the rearrangement products on a timescale faster than equilibration, leads to the premise of stereoselective nucleophilic trapping reactions. In effect, as the nucleophilic trapping is competitive with rearrangement via an ordered contact ion pair, then the order in the ion pair should serve as a stereodirecting element in the nucleophilic trapping reaction, thereby providing the basis for a stereochemical memory effect. The possibility of stereochemical memory effects of this kind marks a fundamental difference between alkene radical cations generated in the confines of a contact ion pair by the fragmentation method, and those generated in the more classical sense by one-electron oxidation of alkenes.

In a variation on the theme of diastereoselective trapping by alcohols, two stereoisomeric precursors of a common alkene radical cation were found to give different product ratios (Scheme 19) [134]. While the fact that both substrates give the same major isomer of the product establishes an important role of the methyl stereogenic center in directing this reaction, the different product ratios demand that the counterion be taken into consideration. In the case of the more selective reaction, the directing effect of the methyl stereogenic center is enhanced by nucleophilic attack on the same face as the departing phosphate. In the less selective case, the stereochemical memory effect works against the directing effect of the benzylic stereogenic center. Overall, the stereochemical memory effect due to the contact ion pair favors nucleophilic attack on the same face of the system from which the phosphate has departed. This is presumably explained in this intermolecular reaction by hydrogen bonding between the departing phosphate and the incoming alcohol. This type of selectivity recalls that seen in some closed-shell contact ion pair reactions, wherein the nucleophile often is incorporated on the same face of the cation from which the leaving group departed [82, 83].

In a more complex elaboration of the hydrogen atom abstraction/nucleophilic cyclization route to tetrahydrofurans (Scheme 20), a carbohydrate-based *N*-alkoxy phthalimide was converted to a spirocyclic acetal in excellent yield and diastereoselectivity (Scheme 33) [136]. In this cyclization, nucleophilic attack takes place from the *endo* face of the trioxabicyclo[3.3.0]octane

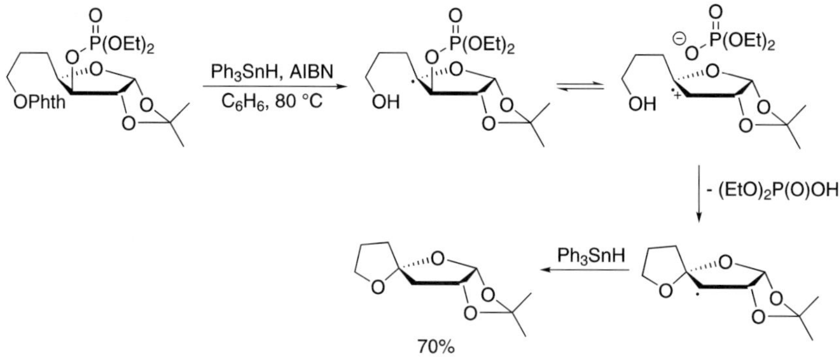

Scheme 33 Conversion of a carbohydrate-based *N*-alkoxy phthalimide to a spirocyclic acetal

skeleton on the opposite face of the alkene radical cation to the one shielded by the just-departed phosphate [136]. A closely related cyclization has also been described in a nucleotide-based system [112].

The reactions of Schemes 19 and 33 differ fundamentally in so far as one employs the phosphate to direct the incoming nucleophile by means of hydrogen bonding, whereas in the other, the phosphate serves as a steric shield to one side of the alkene radical cation. Presumably, this dichotomy can be rationalized in terms of the high effective molarity of the intramolecular nucleophile overcoming any need for hydrogen bonding, resulting in rapid cyclization on the opposite face of the alkene radical cation to that shielded by the bulky phosphate.

In an extensive investigation of the stereochemical memory effect, a series of six diastereomeric pairs of substrates was prepared to probe the effect of single, then multiple substituents on the 5-*exo* cyclization of amines onto alkene radical cations [144, 145]. Overall, these cyclizations were highly diastereoselective and were accounted for by a transition-state model employing a chairlike transition state with attack of the nucleophilic amine on the opposite face of the alkene radical to the one shielded by the phosphate anion in the initial contact ion pair (Scheme 34), as exemplified in Schemes 35 and 36.

The tricyclic product obtained in the example of Scheme 36 arises by an oxidative radical cyclization onto the benzyl group following the nucleophilic ring closure. Interestingly, only one diastereomer of this product is formed regardless of the substrate configuration. This was attributed to the need for a specific steric buttressing interaction with an adjacent methyl group forcing the benzylic group into the proximity of the ring-closed radical.

The transition-state model for these cyclizations (Scheme 34) differs fundamentally from the well-established Beckwith–Houk transition model for radical cyclizations [130, 146–148]. Thus, while both models invoke chairlike transition states, without excluding the possibility of twist boatlike systems in some instances, the Beckwith–Houk model involves full conformational

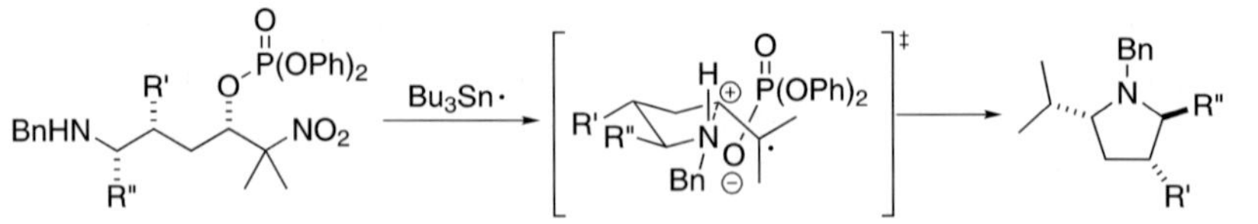

Scheme 34 Chairlike transition-state model for cyclization in a contact ion pair

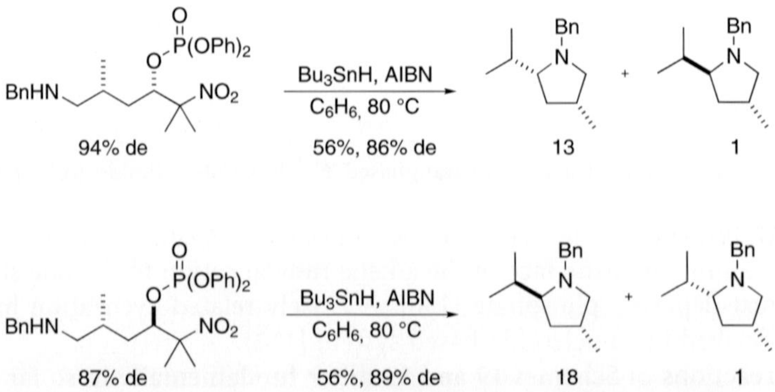

Scheme 35 Diastereoselective cyclization of a pair of diastereomeric substrates

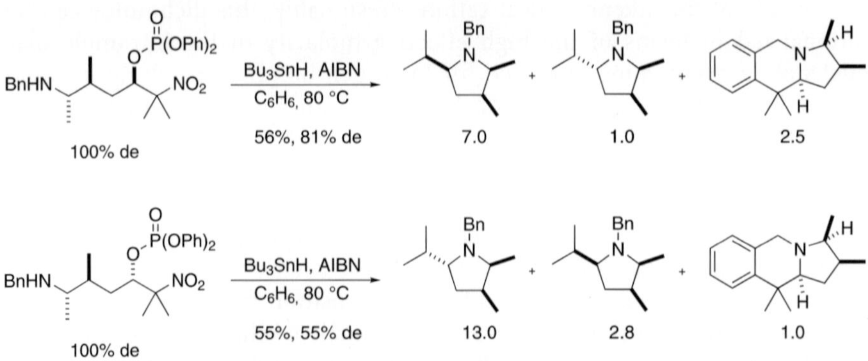

Scheme 36 Diastereoselective cyclization of a further pair of diastereomeric amines

equilibration of the system before cyclization. In contrast, the model presented in Scheme 34 necessitates cyclization before equilibration of the components on the contact radical cation/anion pair if there is to be a stereochemical memory effect. For the same reason, the model differs from those proposed by Moeller and Mattay for the cyclizations of alkene radical cations generated by one-electron oxidation of alkenes [149, 150].

An exception to the general model arises for systems bearing a substituent directly adjacent to the alkene radical cation. Here, the *syn* diastereomer cyclizes with a high degree of stereocontrol, as predicted by the model (Scheme 37) [131, 144, 145].

Conversely, the *anti* isomer cyclizes with much lower selectivity in the opposite sense to that predicted by the model (Scheme 38). This situation arises because cyclization on the initial contact ion pair, on the opposite face to

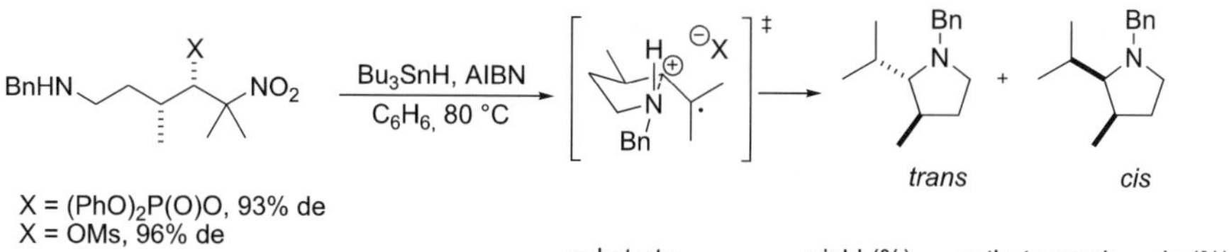

substrate	yield (%)	ratio *trans*:*cis*	de (%)
X = $(PhO)_2P(O)O$	50	>25:1	>95
X = OMs	54	>25:1	>95

Scheme 37 Favorable effect of an adjacent substituent on cyclization stereochemistry

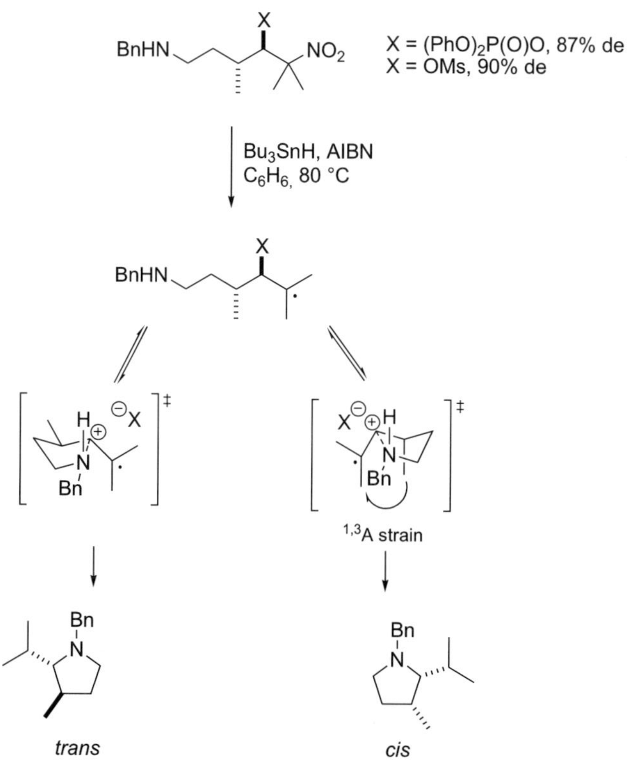

substrate	yield (%)	ratio *trans*:*cis*	de (%)
X = $(PhO)_2P(O)O$	45	>2.4:1	40
X = OMs	47	>9.0:1	80

Scheme 38 Unfavorable effect of an adjacent substituent on cyclization stereochemistry

that shielded by the leaving group, engenders a high degree of ^{1,3}A-strain in the alkene radical cation. This raises the barrier for cyclization according to the model of Scheme 34, allowing time for the system to equilibrate and select the alternative chairlike transition state which minimizes steric interactions [131, 144, 145]. This situation corresponds much more closely to the equilibrated models of Beckwith and Houk, and of Moeller and Mattay, for radical and radical cation cyclizations, respectively. It is noteworthy that in the case of the *syn* isomer there is little difference between the use of diphenyl phosphate and mesylate leaving groups, but the mesylate allows a greater degree of equilibration in the case of the *anti* isomer.

13
Enantioselectivity in Nucleophilic Cyclizations

True stereochemical memory effects, lacking complications from additional stereogenic centers, are observable in simple enantiomerically enriched β-nitroalkyl phosphates. With an alcohol as nucleophile and diphenyl phosphate as leaving group, complete racemization was observed, which was attributed to the low nucleophilicity of the alcohol in comparison to the rate of equilibration of the ion pair (Scheme 39) [141, 142]. The low yield of the anticipated tetrahydrofuran seen with this substrate is due to the formation of a 1,3,2-dioxaphosphepane which results from nucleophilic attack at the phosphate by the alcohol. This type of side reaction is often problematic when attempting cyclizations with alcohols, and was the exclusive reaction pathway for substrates carrying cyclization-enhancing *gem*-dialkyl groups [142].

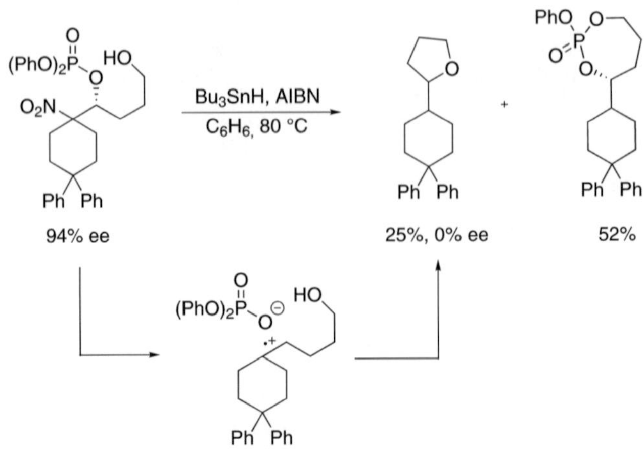

Scheme 39 Racemization in the cyclization of a β-nitroalkyl phosphate with an alcohol as nucleophile

The contrast between the lack of enantioselectivity in Scheme 39 and the moderate to excellent diastereoselectivity seen with alcohol nucleophiles in Schemes 19 and 33 can be attributed to the difference in leaving groups (diphenyl phosphate vs diethyl phosphate) and to the differences in the radical cations themselves, all of which impinge on the rate of equilibration of the contact alkene radical cation/anion pair.

With amines as nucleophiles attack at the alkene radical cation is more rapid and the problem of substitution at phosphorus much less severe, such that enantioselective formation of pyrrolidines becomes possible (Scheme 40) [141, 142]. Interestingly, neither the exchange of the diphenyl phosphate group for its diethyl counterpart, nor the inclusion of a *gem*-dialkyl group had a significant influence on the enantioselectivity of these processes, indicative of an extremely rapid cyclization. (Significant *gem*-dialkyl effects in the formation of seven-membered rings by alkene radical cation cyclizations conducted under electrochemical conditions have been observed [18].) Likewise the enantioselective formation of a piperidine ring system reflects the high rate of ring closure (Scheme 41) [141, 142].

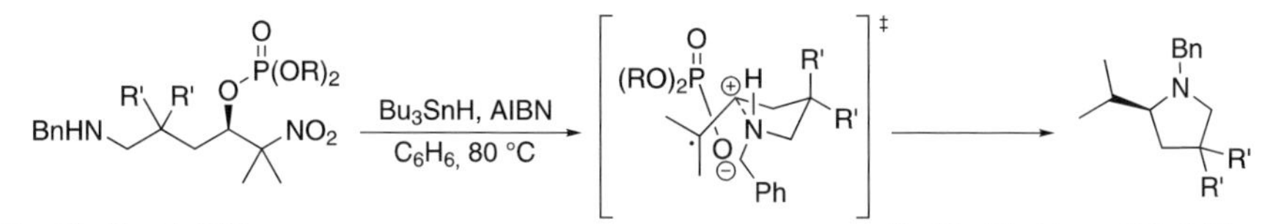

R = Ph, R' = H, 85% ee
R = Ph, R' = Bn, 79% ee
R = Et, R' = H, 85% ee

R = Ph, R' = H 43%, 61% ee
R = Ph, R' = Bn 44%, 60% ee
R = Et, R' = H 45%, 62% ee

Scheme 40 Enantioselective formation of pyrrolidines

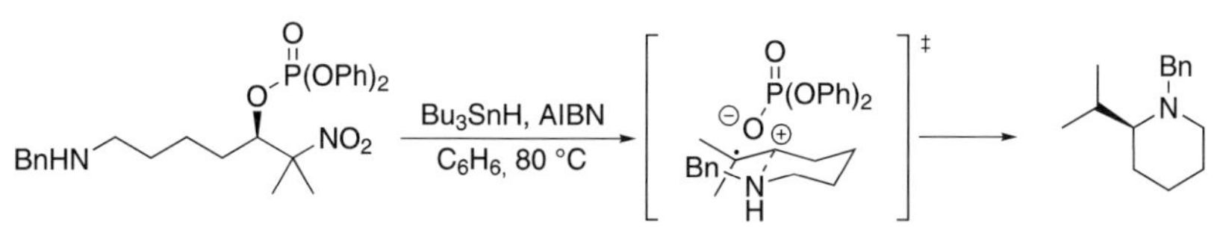

86% ee

41%, 60% ee

Scheme 41 Enantioselective formation of a piperidine ring system

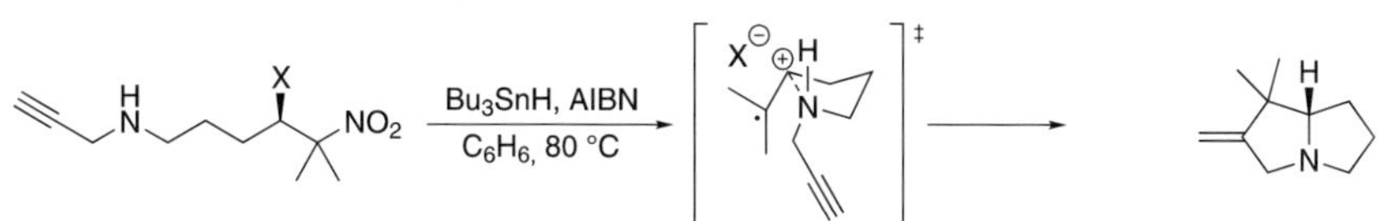

X = OMs, 93% ee
X = (PhO)$_2$P(O)O, 85% ee

X = OMs 60%, 80% ee
X = (PhO)$_2$P(O)O 64%, 60% ee

Scheme 42 Enantioselective cyclization in tandem processes leading to bicyclic systems

When the nucleophile bears an appropriately unsaturated chain, these enantioselective cyclizations can be used to advantage in tandem processes leading to bicyclic systems (Scheme 42) [131, 141, 142]. The greater enantioselectivity observed with the mesylate group in this example may be due to the lower degree of stabilization of the alkene radical cation in the looser ion pair, which leads to more rapid cyclization.

14 Miscellaneous

A series of *N*-allyl sulfamates, phosphoramides, and phosphorimidates was prepared to explore the possibility of O→N rearrangements via the intermediacy of the contact alkene radical cation/anion pair, followed by 5-*exo*-trigonal radical cyclizations (Fig. 4) [142].

Unfortunately, while fragmentation was observed for the sulfamate and the phosphorimidate, the desired rearrangement did not take place. In the case of the phosphoramide, rearrangement did occur but unfortunately through the formal 1,2-pathway common in simple phosphates, and not by the desired O→N shift (Scheme 43). Interestingly, the formation of an eight-membered ring, by an 8-*endo*-trig process, was observed as a minor competing pathway with this substrate, suggesting that the phosphoramide is a relatively poor leaving group.

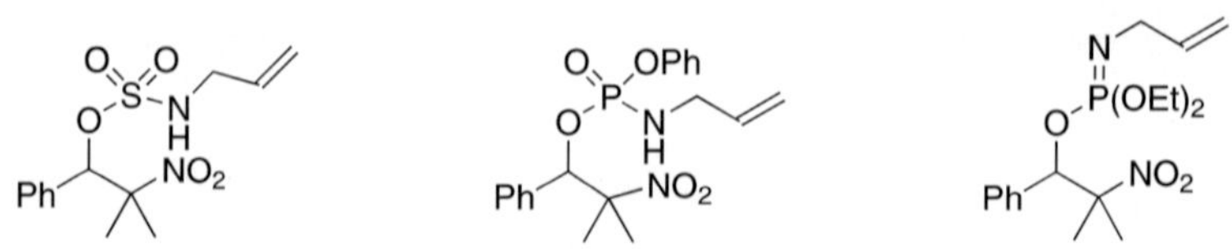

Fig. 4 Sulfamates, phosphoramides, and phosphorimidates studied

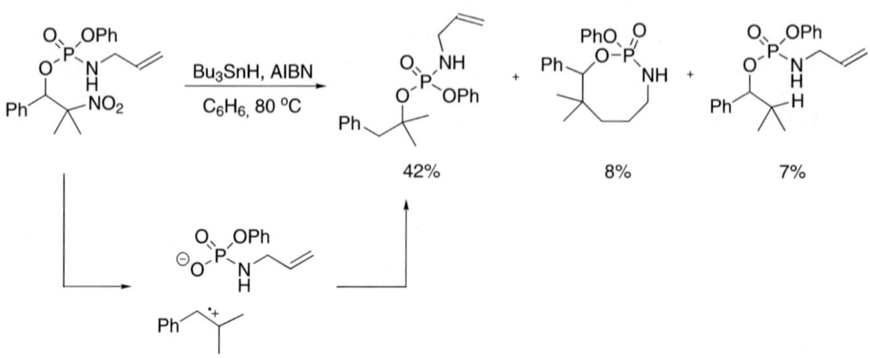

Scheme 43 Phosphoramide rearrangement

Acknowledgement We thank the Ministère des Affaires Étrangères, France, for a Lavoisier Fellowship in support of FB, the National Science Foundation for support of our work in this area (CHE 9986200), the numerous students in the Crich group who contributed significantly to our work in this area, whose names are listed in the references, and Professors Martin Newcomb and Athel Beckwith for stimulating collaborations and frequent exchanges of ideas.

References

1. Tanko JM, Phillips JP (1999) J Am Chem Soc 121:6078
2. Hintz S, Heidbreder A, Mattay J (1996) Top Curr Chem 177:77
3. Bauld NL, Bellville DJ, Harirchian B, Lorenz KT, Pabon RA, Reynolds DW, Wirth DD, Chiou H-S, Marsh BK (1987) Acc Chem Res 20:371
4. Schmittel M, Burghart A (1997) Angew Chem Int Ed 36:2550
5. Mangion D, Arnold DR (2002) Acc Chem Res 35:297
6. Dalko PI (1995) Tetrahedron 51:7579
7. Moeller KD (2000) Tetrahedron 56:9527
8. Little RD, Moeller KD (2002) Electrochem Soc Interface 11:36
9. Bauld NL, Roh Y (2002) Curr Org Chem 6:647
10. Nair V, Balagopal L, Rajan R, Mathew J (2004) Acc Chem Res 37:21
11. Ogibin YN, Nikishin GI (2001) Russ Chem Rev 70:543
12. Todres ZV (2003) Organic radical ions: chemistry and applications. Dekker, New York
13. Oelgemöller M, Bunte JO, Mattay J (2005) Mol Supramol Photochem 12:269
14. Wiest O, Oxgaard J, Saettel NJ, Adv Phys Org Chem 38:87
15. Moeller KD (1997) Top Curr Chem 185:49
16. Tabakovic I (1997) Top Curr Chem 185:87
17. Clark T (1996) Top Curr Chem 177:1
18. Sperry JB, Wright DL (2005) J Am Chem Soc 127:8034
19. Sperry JB, Constanzo JR, Jasinski J, Butcher RJ, Wright DL (2005) Tetrahedron Lett 46:2789
20. Sperry JB, Wright DL (2005) Tetrahedron Lett 46:411
21. Mihelcic J, Moeller KD (2004) J Am Chem Soc 126:9106
22. Huang Y-T, Moeller KD (2004) Org Lett 6:4199
23. Brandt JD, Moeller KD (2005) Tetrahedron Lett 7:3553
24. Fürmeier S, Metzger JO (2004) J Am Chem Soc 126:14485
25. Meyer S, Koch R, Metzger JO (2003) Angew Chem Int Ed 42:4700
26. Chiba K, Miura T, Kim S, Kitano Y, Tada M (2001) J Am Chem Soc 123:11314
27. Grützmacher H-F (2003) Curr Org Chem 7:1565
28. Bauld NL (1997) Radicals, radical ions, and triplets: the spin-bearing intermediates of organic chemistry. Wiley, New York
29. Rinderhagen H, Mattay J (2004) Chem Eur J 10:851
30. Bunte JO, Mattay J (2004) In: Horspool W, Lenci F (eds) CRC handbook of organic photochemistry and photobiology. CRC, Boca Raton, p 10/1
31. Bauld NL, Gao D (2001) In: Balzani V (ed) Electron transfer in chemistry and biochemistry, vol 2. Wiley, Weinheim, p 133
32. Gilbert BC, Larkin JP, Norman ROC (1972) J Chem Soc Perkin Trans 2:794
33. Behrens G, Koltzenburg G, Schulte-Frohlinde D (1982) Z Naturforsch 37c:1205

34. von Sonntag C (1987) The chemical basis of radiation biology. Taylor and Francis, London
35. von Sonntag C, Hagen U, Schon-Bopp A, Schulte-Frohlinde D (1981) In: Lett JT, Adler H (eds) Advances in radiation biology, vol 9. Academic, New York, p 109
36. von Sonntag C (1980) Adv Carbohydr Chem Biochem 37:7
37. Chatgilialoglu C, O'Neill P (2001) Exp Gerontol 36:1459
38. Glatthar R, Spichty M, Gugger A, Batra R, Damm W, Mohr M, Zipse H, Giese B (2000) Tetrahedron 56:4117
39. Pogozelski WK, Tullius TD (1998) Chem Rev 98:1089
40. Breen AP, Murphy JA (1995) Free Radic Biol Med 18:1033
41. Knorre DG, Fedorova OS, Frolova EI (1993) Russ Chem Rev 62:65
42. Pratviel G, Bernadou J, Meunier B (1995) Angew Chem Int Ed 34:746
43. Stubbe J, Kozarich JW, Wu W, Vanderwall DE (1996) Acc Chem Res 29:322
44. Dedon PC, Goldberg IH (1992) Chem Res Toxicol 5:311
45. Hecht SM (2000) J Nat Prod 63:158
46. Giese B (2000) Acc Chem Res 33:631
47. Giese B (2004) Top Curr Chem 236:27
48. Surzur J-M, Teissier P (1967) CR Acad Sci 264:1981
49. Crich D, Yao Q (1993) J Am Chem Soc 115:1165
50. Koch A, Lamberth C, Wetterich F, Giese B (1993) J Org Chem 58:1083
51. Beckwith ALJ, Crich D, Duggan PJ, Yao Q (1997) Chem Rev 97:3273
52. Crich D (2001) In: Renaud P, Sibi M (eds) Radicals in organic synthesis, vol 2. Wiley, Weinheim, p 188
53. Clark T, Nelsen SF (1988) J Am Chem Soc 110:868
54. Takahashi O, Kikuchi O (1994) J Mol Struct (Theochem) 313:207
55. Liu Y-J, Huang M-B (2001) J Mol Struct (Theochem) 536:133
56. Abrams Ml, Valeev EF, Sherrill CD, Crawford TD (2002) J Phys Chem A 106:2671
57. Salhi-Benachenhou N, Engels B, Huang M-B, Lunell S (1998) Chem Phys 236:53
58. Kochi JK, Rathore R, Zhu J, Lindeman SV (2000) Angew Chem Int Ed 39:3671
59. Nelsen SF, Reinhardt LA, Tran HQ, Clark T, Chen G-F, Pappas RS, Williams F (2002) Chem Eur J 8:1074
60. Gerson F, Lopez J, Krebs A, Rüger W (1981) Angew Chem Int Ed 20:95
61. Behrens G, Bothe E, Koltzenberg G, Schulte-Frohlinde D (1980) J Chem Soc Perkin Trans 2:883
62. Behrens G, Koltzenburg G, Ritter A, Schulte-Frohlinde D (1978) Int J Radiat Biol 33:163
63. Koltzenburg G, Behrens G, Schulte-Frohlinde D (1982) J Am Chem Soc 104:7311
64. Müller SN, Batra R, Senn M, Giese B, Kisel M, Shadyro O (1997) J Am Chem Soc 119:2796
65. Giese B, Burger J, Kang TW, Kesselheim C, Wittmer T (1992) J Am Chem Soc 114:7322
66. Giese B, Beyrich-Graf X, Burger J, Kesselheim C, Senn M, Schafer T (1993) Angew Chem Int Ed 32:1742
67. Gugger A, Batra R, Rzadek P, Rist G, Giese B (1997) J Am Chem Soc 119:8740
68. Lancelot SF, Cozens FL, Schepp NP (2003) Org Biomol Chem 1:1972
69. Crich D, Suk D-H, Hao X (2002) Tetrahedron 58:5789
70. Crich D, Jiao X-Y (1996) J Am Chem Soc 118:6666
71. Beckwith ALJ, Duggan PJ (1996) J Am Chem Soc 118:12838
72. Beckwith ALJ, Thomas CB (1973) J Chem Soc Perkin Trans 2:861
73. Lacote E, Renaud P (1998) Angew Chem Int Ed 37:2259

74. Choi S-Y, Crich D, Horner JH, Huang X, Newcomb M, Whitted PO (1999) Tetrahedron 55:3317
75. Reichardt C (1994) Chem Rev 94:2319
76. Whitted PO, Horner JH, Newcomb M, Huang X, Crich D (1999) Org Lett 1:153
77. Newcomb M, Horner JH, Whitted PO, Crich D, Huang X, Yao Q, Zipse H (1999) J Am Chem Soc 121:10685
78. Bagnol L, Horner JH, Newcomb M (2003) Org Lett 5:5055
79. Bales BC, Horner JH, Huang X, Newcomb M, Crich D, Greenberg MM (2001) J Am Chem Soc 123:3623
80. Sprecher M (1994) Chemtracts 7:115
81. Winstein S, Clippinger E, Fainberg AH, Heck R, Robinson GC (1956) J Am Chem Soc 78:328
82. Richard JP, Amyes Tl, Toteva MM, Tsuji Y (2004) Adv Phys Org Chem 39:1
83. Raber DJ, Harris JM, Schleyer PvR (1974) In: Szwarc M (ed) Ions and ion pairs in organic reactions, vol 2. Wiley, New York, p 247
84. Horner JH, Bagnol L, Newcomb M (2004) J Am Chem Soc 126:14979
85. Taxil E, Bagnol L, Horner JH, Newcomb M (2003) Org Lett 5:827
86. Horner JH, Taxil E, Newcomb M (2002) J Am Chem Soc 124:5402
87. Newcomb M, Miranda N, Huang X, Crich D (2000) J Am Chem Soc 122:6128
88. Arnold BR, Noukakis D, Farid S, Goodman JL, Gould IR (1995) J Am Chem Soc 117:4399
89. Newcomb M (1993) Tetrahedron 49:1151
90. Peukert S, Batra R, Giese B (1997) Tetrahedron Lett 38:3507
91. Schepp NP, Shukla D, Bauld NL, Johnston LJ (1997) J Am Chem Soc 119:10325
92. Lew CSQ, Brisson JR, Johnston LJ (1997) J Org Chem 62:4047
93. Behrens G, Bothe E, Koltzenburg G, Schulte-Frohlinde D (1981) J Chem Soc Perkin Trans 2:143
94. Johnston LJ, Schepp NP (1993) J Am Chem Soc 115:6564
95. Johnston LJ, Schepp NP (1996) Adv Electron Transfer Chem 5:41
96. Johnston LJ, Schepp NP (1995) Pure Appl Chem 67:71
97. Saebo S, Beckwith ALJ, Radom L (1984) J Am Chem Soc 106:5119
98. Zipse H (2003) Adv Phys Org Chem 38:111
99. Barclay LRC, Griller D, Ingold KU (1982) J Am Chem Soc 104:4399
100. Wang Y, Grimme S, Zipse H (2004) J Phys Chem A 108:2324
101. Zipse H, Bootz M (2001) J Chem Soc Perkin Trans 2:1566
102. Zipse H (1994) Angew Chem Int Ed 33:1985
103. Zipse H (1999) Acc Chem Res 32:571
104. Peukert S, Giese B (1996) Tetrahedron Lett 37:4365
105. Clive DLJ, Chittattu GJ, Farina V, Kiel WA, Menchen SM, Russell CG, Singh A, Wong CK, Curtis N (1980) J Am Chem Soc 102:4438
106. Beckwith ALJ, Pigou PE (1986) Aus J Chem 39:77
107. Filzen GF (1996) PhD thesis, University of Illinois at Chicago
108. Crich D (1987) Aldrichim Acta 20:35
109. Crich D, Quintero L (1989) Chem Rev 89:1413
110. Motherwell WB, Imboden C (2001) In: Renaud P, Sibi MP (eds) Radicals in organic synthesis, vol 1. Wiley, Weinheim, p 109
111. Crich D, Huang X, Newcomb M (2000) J Org Chem 65:523
112. Crich D, Huang W (2001) J Am Chem Soc 123:9239
113. Ono N, Miyake H, Kamimura A, Hamamoto I, Tamura R, Kaji A (1985) Tetrahedron 41:4013

114. Ono N (2001) The nitro group in organic synthesis. Wiley, New York
115. Giese B, Groninger KS (1990) Org Synth 69:66
116. Giese B, Gilges S, Groninger KS, Lamberth C, Witzel T (1988) Liebigs Ann Chem 615
117. Quiclet-Sire B, Zard SZ (1996) J Am Chem Soc 118:9190
118. Korth H-G, Sustmann R, Groninger KS, Leisung M, Giese B (1988) J Org Chem 53:4364
119. Crich D, Yao Q, Filzen GF (1995) J Am Chem Soc 117:11455
120. Kocovsky P, Stary I, Turecek F (1986) Tetrahedron Lett 27:1513
121. Beckwith ALJ, Duggan PJ (1992) J Chem Soc Perkin Trans 2:1777
122. Beckwith ALJ, Duggan PJ (1998) Tetrahedron 54:4623
123. Beckwith ALJ, Duggan PJ (1998) Tetrahedron 54:6919
124. Crich D, Beckwith ALJ, Filzen GF, Longmore RW (1996) J Am Chem Soc 118:7422
125. Crich D, Huang X, Beckwith ALJ (1999) J Org Chem 64:1762
126. Crich D, Sartillo-Piscil F, Quintero-Cortes L, Wink DJ (2002) J Org Chem 67:3360
127. Crich D, Suk D-H (2004) Can J Chem 82:75
128. Crich D, Suk D-H, Sun S (2003) Tetrahedron Asymmetry 14:2861
129. Clive DLJ (1975) J Chem Soc Chem Commun: 353
130. Beckwith ALJ (1981) Tetrahedron 37:3073
131. Suk D-H (2004) PhD thesis, University of Illinois at Chicago
132. Koch A, Giese B (1993) Helv Chim Acta 76:1687
133. Walling C, Cioffari A (1972) J Am Chem Soc 94:6064
134. Crich D, Gastaldi S (1998) Tetrahedron Lett 39:9377
135. Crich D, Huang X, Newcomb M (1999) Org Lett 1:225
136. Sartillo-Piscil F, Vargas M, Anaya de Parrodi C, Quintero L (2003) Tetrahedron Lett 44:3919
137. Kim S, Lee TA, Song Y (1998) Synlett 501
138. Kabasakalian P, Townley ER, Yudis MD (1962) J Am Chem Soc 84:2716
139. Crich D, Ranganathan K, Neelamkavil S, Huang X (2003) J Am Chem Soc 125:7942
140. Crich D, Neelamkavil S (2002) Org Lett 4:2573
141. Crich D, Shirai M, Rumthao S (2003) Org Lett 5:3767
142. Crich D, Shirai M, Brebion F, Rumthao S (2006) Tetrahedron (in press)
143. Crich D, Ranganathan K, Huang X (2001) Org Lett 3:1917
144. Crich D, Ranganathan K (2002) J Am Chem Soc 124:12422
145. Crich D, Ranganathan K (2005) J Am Chem Soc 127:9924
146. Beckwith ALJ, Schiesser CH (1985) Tetrahedron 41:3925
147. Spellmeyer DC, Houk KN (1987) J Org Chem 52:959
148. Curran DP, Porter NA, Giese B (1996) Stereochemistry of radical reactions. Wiley, Weinheim
149. Liu B, Duan S, Sutterer AC, Moeller KD (2002) J Am Chem Soc 124:10101
150. Bunte JO, Heilmann EK, Hein B, Mattay J (2004) Eur J Org Chem 3535

Top Curr Chem (2006) 263: 39–69
DOI 10.1007/128_020

Published online: 13 December 2005

The Mechanism of Epoxide Opening through Electron Transfer: Experiment and Theory in Concert

Kim Daasbjerg[1] (✉) · Heidi Svith[1] · Stefan Grimme[2] (✉) · Mareike Gerenkamp[2] · Christian Mück-Lichtenfeld[2] · Andreas Gansäuer[3] (✉) · Andriy Barchuk[3]

[1]Department of Chemistry, University of Aarhus, Langelandsgade 140, 8000 Aarhus C, Denmark
kdaa@chem.au.dk

[2]Organisch-Chemisches Institut, Westfälische Wilhelms-Universität, Correnstr. 40, 49149 Münster, Germany
grimmes@uni-muenster.de

[3]Kekulé Institut für Organische Chemie und Biochemie, Universität Bonn, Gerhard-Domagk-Str. 1, 53121 Bonn, Germany
andreas.gansaeuer@uni-bonn.de

Abstract This review gives a description of the mechanism of reductive epoxide opening through single-electron transfer. A number of electron-transfer reagents are compared and the most promising titanocene complexes are studied in detail. The mechanism of epoxide opening was established by cyclic voltammetry, kinetic measurements, DFT calculations, and synthetic studies. The results are used to devise more selective reagents.

Keywords Catalysis · Cyclic voltammetry · DFT calculations · Electron transfer · Radicals

Abbreviations

B-P functional	Becke–Perdew functional
Coll	2,4,6-Collidine
Cp	Cyclopentadienyl
CV	Cyclic voltammetry
DFT	Density functional theory
DMF	*N*,*N*-Dimethylformamide
dr	Diastereomeric ratio
equiv.	Equivalents
ET	Electron transfer
EtOAc	Ethyl acetate
THF	Tetrahydrofuran
TZVP	Triple-zeta valence polarization

1 Introduction

Epoxides are amongst the most frequently employed substrates in organic synthesis. This is due to the ease of their preparation from readily available precursors, e.g., olefins and carbonyl compounds [1, 2] and their high reactivity [3]. The latter point arises mainly from the strain inherent in the three-membered ring that is released during ring opening. Epoxides, especially when prepared in high enantiomeric excess, have been very useful in S_N2 reactions in this respect. An alternative approach to exploiting the high reactivity of the strained epoxide is constituted by ring-opening reactions utilizing electron-transfer reagents [4, 5]. In this review we highlight the mechanistic aspects, from establishing experimental criteria for this reactivity to studies of catalyst composition by cyclic voltammetry and structural investigations of pertinent intermediates and transition structures by computational chemistry. Key synthetic perspectives arising from these studies will be briefly outlined.

1.1 Epoxide Opening under Birch and Modified Birch Conditions

In the light of the success of the Birch conditions for reducing organic compounds it is not surprising that epoxides can be opened by solvated electrons [6–9]. The initially formed radical is then further reduced to give carbanionic species, which do not display the reactivity of radicals. This concept has been extended by Bartmann [10], Cohen et al. [11], Conrow [12], and Yus et al. [13, 14] who employed aromatic radical anions as the reduc-

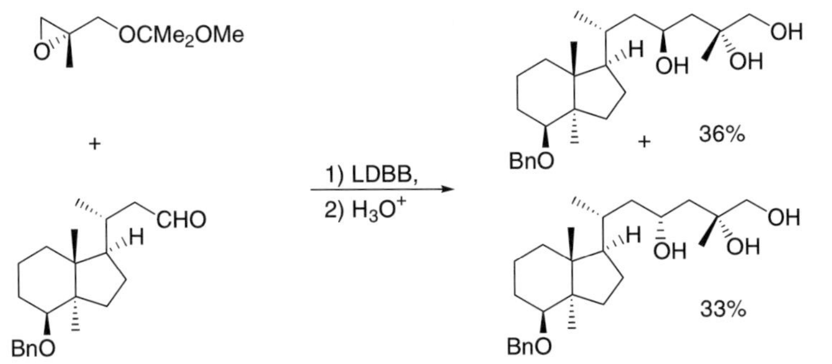

Scheme 1 Modified steroids from functionalized organolithiums [12]

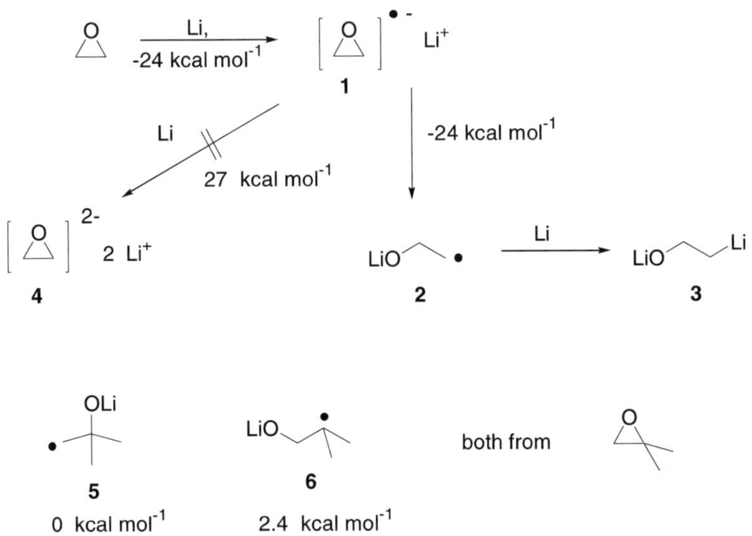

Scheme 2 Mechanism of epoxide opening through direct ET

ing agents in many synthetically useful applications. An example is shown in Scheme 1.

Radical formation was investigated by Cohen et al. [11], and their findings are summarized in Scheme 2. The most important result was the identification of the lithiated radical anion **1** as decisive intermediate. Opening of **1** was exothermic by about 24 kcal mol^{-1} and produced radical **2** which was reduced to **3**. Dianion **4** was not involved.

The calculations also suggested that **5** was favored over **6** by 2.4 kcal mol^{-1} as found experimentally. The explanation was based on the higher stability of a tertiary alkoxide compared to a primary alkoxide [15], which outweighed the opposite trend for radical stabilization. Epoxide opening was irreversible [16].

2
Epoxide Opening by Low-Valent Metal Complexes

With respect to the utilization of the intermediate radicals for organic synthesis, the use of low-valent metal complexes is more promising. In this manner the advantages of Lewis acid catalysis and radical chemistry can be combined. This is achieved by activating the epoxide toward ET by complexation with the metal and controlling the regio- and stereoselectivity of epoxide opening through the metal and its ligands. Of course, epoxide activation through S_N-type reactions and succeeding reduction of metalated halohydrins must be avoided.

The general idea of this concept was first outlined by Nugent and RajanBabu [17–20] as shown in Scheme 3, and constitutes an analogue of the well-established opening of a cyclopropylcarbinyl radical [21,22]. Titanocenes have emerged as the most powerful reagents in these transformations. However, it is clearly attractive to find other metal complexes in order to develop novel reactivity patterns.

2.1
Experimental Evidence for the Formation of *β*-Metaloxy Radicals

For this purpose, and to establish criteria for judging their performance, ET reagents that do not react through two-electron processes via oxidative insertions or S_N2-type reactions [23–25] have been studied in three mechanistically relevant reactions [26]. It is generally agreed that with ET reagents the pivotal *β*-metaloxy radicals are indeed formed if the deoxygenation reaction of epoxides proceeds with low stereoselectivity to mixtures of the corresponding (*E*) and (*Z*) olefins as shown in Scheme 4 [17–20]. This argument is based on the nonselective trapping of the configurationally labile radical intermediate by the second equivalent of the ET reagent. The resulting diastereomeric mixture undergoes elimination to give the depicted mixture of olefins.

The deoxygenation of simple unfunctionalized epoxides has already been investigated with titanocene [17–20] and samarium [27] reagents. Usually both metal complexes give mixtures of the isomers with low selectivity. Epoxide 7 investigated here is mechanistically more interesting because the

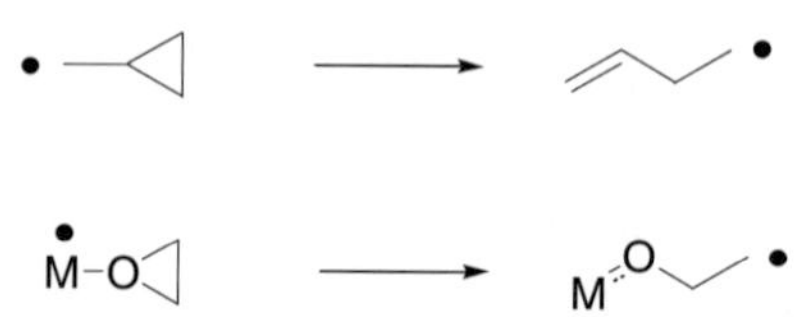

Scheme 3 Formal analogy between epoxide opening by metal complexes and cyclopropylcarbinyl radical opening

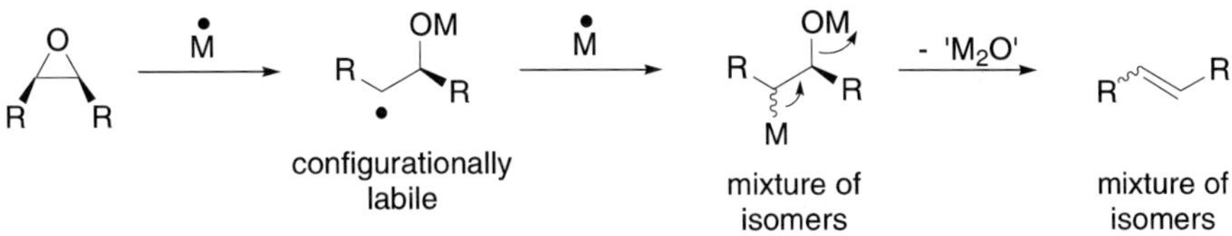

Scheme 4 Epoxide deoxygenation via ET from metal complexes

organometallic intermediate formed after reductive trapping with a second equivalent of the low-valent metal complex can give two different elimination products. Consequently, the issue of regioselectivity of the overall transformation and the factors controlling it are raised, as shown in Scheme 5.

We decided to compare SmI_2, $CrCl_2$, $[V_2Cl_3(THF)_6]_2[Zn_2Cl_6]$, obtained by reduction of $VCl_3(THF)_3$ [28] with Zn dust in THF, and Zn- or Mn-reduced solutions of Cp_2TiCl_2 [17–20] as reducing agents. The vanadium-based reagent has to the best of our knowledge not yet been used in epoxide openings. It was proven to give excellent results in pinacol-type reactions introduced by Pedersen et al. [29–33]. $CrCl_2$ has, to the best of our knowledge, only been used in deoxygenation reactions of cyclohexene and styrene oxide where no problems of selectivity could occur [34]. Our results are summarized in Table 1.

Interestingly, $[V_2Cl_3(THF)_6]_2[Zn_2Cl_6]$ and SmI_2 react with complete, albeit opposite, selectivity. Whereas mixtures of the (*E*) and (*Z*) isomers (50 : 50) of **8** are formed in 73% yield with $[V_2Cl_3(THF)_6][Zn_2Cl_6]$, SmI_2 gives only the product of propoxide elimination **9** in 53% yield. No other products were obtained. This rather low mass balance could be due to the noticeable volatility and water solubility of **9**. Surprisingly, the chromium reagent exhibited distinctly lower reactivity even when employed in DMF and in the presence of the diamine ligand ethylene diamine. Besides 49% of reisolated starting material, the deoxygenation products **8** and **9** were obtained in low yields. The titanocene reagent gave **8** in 19% yield as a 58 : 42 mixture of (*E*) and (*Z*) isomers and **9** in 47% yield.

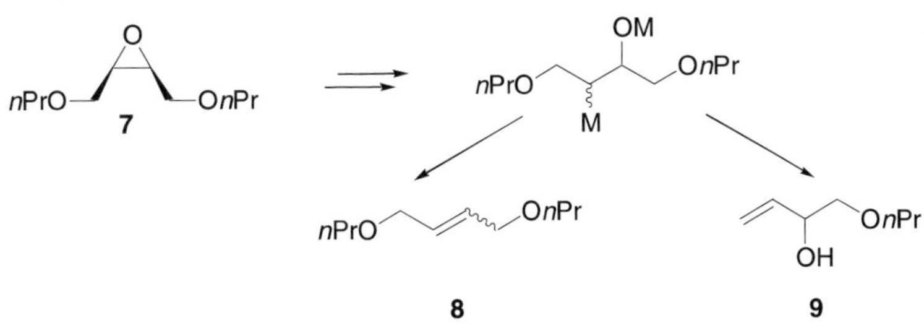

Scheme 5 Possible pathways in chemoselective deoxygenation of **7**

Table 1 Regioselectivity in the deoxygenation of 7

Entry	ET reagent	Product	Yield (%)
1	$CrCl_2$	 **8** (*E*) : (*Z*) = 41:59	19[a]
		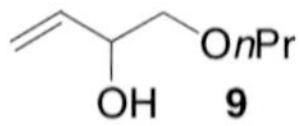 **9**	16
2	$VCl_3(THF)_3$/Zn	 **8** (*E*) : (*Z*) = 50:50	73
3	SmI_2	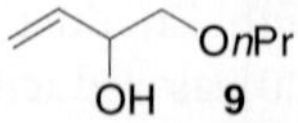 **9**	53
4	Cp_2TiCl	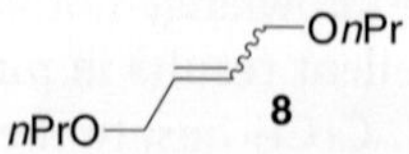 **8** (*E*) : (*Z*) = 58:42	19
		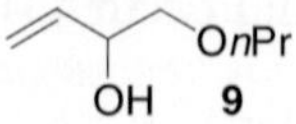 **9**	47

[a]49% substrate reisolated

The deoxygenation of epoxides with the metal complexes mentioned above all seem to proceed via intermediate β-metal oxy radicals. The reaction path after their trapping seems, however, to depend on the Lewis acidity of the ET reagent.

With these results in hand, we turned our attention to the decisive questions on the use of the pivotal β-metal oxy radicals, i.e., the mechanism of their formation and their persistence in a reductive medium. For the latter point it is mandatory to note that radical reactivity for C – H and C – C bond formation can only be exploited if the reduction of the radical is slower than the attempted ensuing radical transformation, e.g., a 5-*exo* cyclization [35]. For SmI_2 it is known that primary alkyl radicals are reduced with rate constants of about 6×10^6 M^{-1} s^{-1} [36–38]. However, 5-*exo* cyclizations of radicals derived from SmI_2-mediated halide abstraction with alkenes and alkynes are well documented and must therefore be considered faster than reductive trapping [35]. The question of the mechanism of radical generation will be addressed by analyzing the chemo- and regioselectivity of epoxide opening and the structure of the products arising from trapping of the intermediates formed.

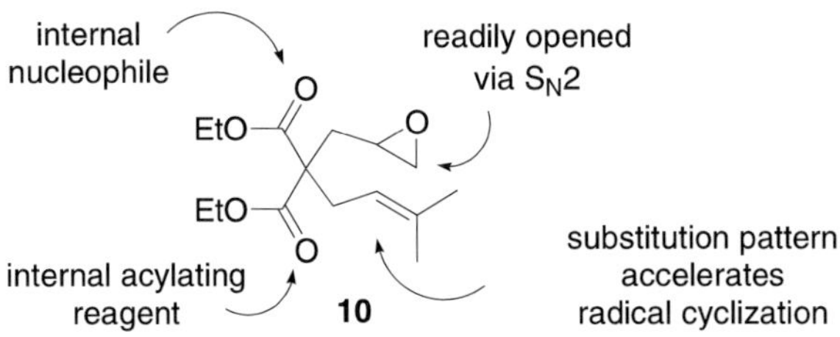

Fig. 1 Structural features of **10** necessary for establishing competing mechanisms of epoxide opening

Therefore, we decided to investigate the ET reagents mentioned above with suitably unsaturated epoxides. Epoxide **10** was chosen as substrate in these reactions for the reasons shown in Fig. 1. Firstly, the ester groups can act as internal nucleophiles for Lewis acid assisted epoxide opening; secondly, the monosubstituted epoxide is readily attacked by external nucleophiles, e.g., iodide; and thirdly, the trisubstituted olefin is known to accelerate 5-*exo* cyclizations and should intercept radicals efficiently [39]. Last but not least, the ester groups can trap alkoxides formed by S_N2 opening of the epoxide as intramolecular acylating reagents. Thus, substrate **10** should allow an analysis of radical generation, trapping, and side reactions. Also, the question of direct epoxide reduction and S_N2 opening followed by reduction of metalated iodohydrins should be resolved by analysis of the opening products.

The results of our investigations based on the opening shown in Scheme 6 are summarized in Table 2. SmI_2 as ET reagent resulted in the formation of two products in good combined yield. Lactone **12** is formed by epoxide opening with iodide via S_N2 reaction and the ensuing fast lactonization. The deoxygenation product **13** is formed by reduction of **12** with 2 equiv. of SmI_2 and β-elimination of a carboxylate. Thus, the epoxide ring is not opened via ET at all and the S_N2 reaction with iodide must be considered much faster. In the case of the vanadium reagent different results were obtained. As expected, no chlorohydrins from external nucleophilic attack could be observed: chloride is only a weak nucleophile. Accordingly, lactone **15** was isolated in low yield (7%). Thus, some vanadium or zinc species was Lewis acidic enough to promote an intramolecular nucleophilic epoxide opening to a small extent. The main product is, however, constituted by the deoxygenation product **14**

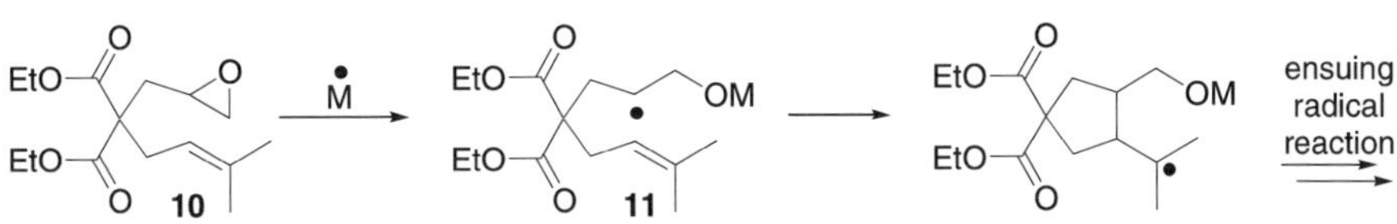

Scheme 6 Radical reactions of **10**

Table 2 Competing pathways in the opening of **10**

Entry	ET reagent	Product	Yield (%)
1	SmI_2	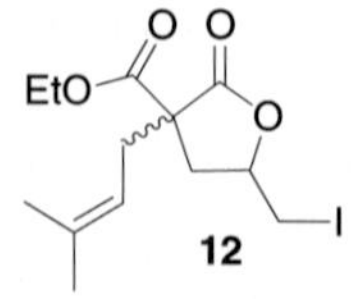 **12**	31, dr = 61:39
		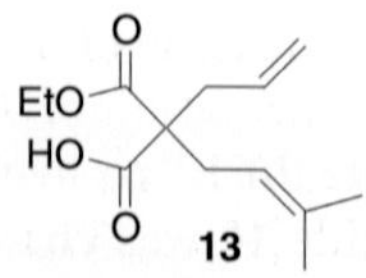 **13**	55
2	$VCl_3(THF)_3$/Zn	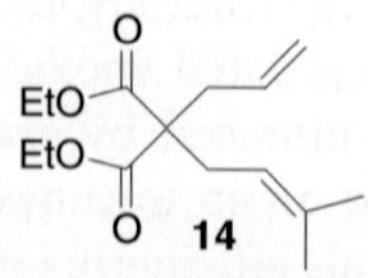 **14**	59
		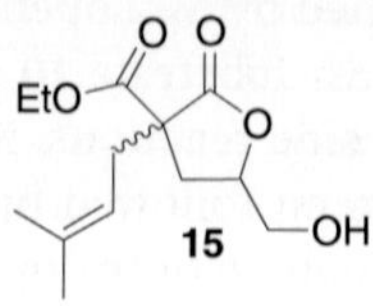 **15**	7 dr = 80:20
3	VCl_3/Zn	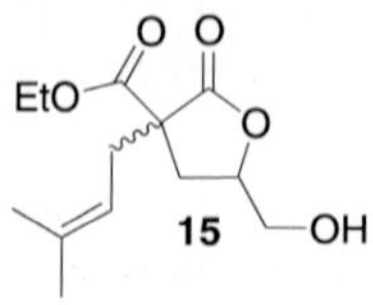 **15**	48 dr = 60:40
4	$VCl_3(THF)_3$	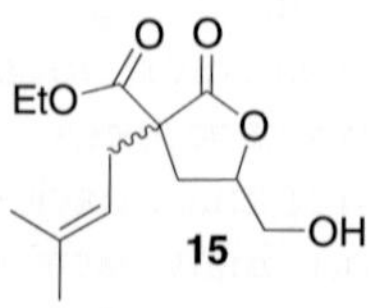 **15**	25 dr = 40:60
5	Cp_2TiCl	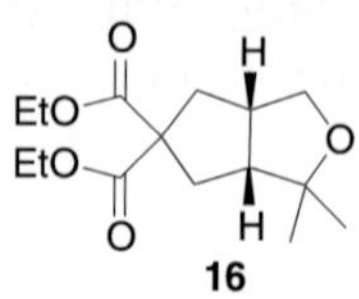 **16**	57

which could be isolated in 59% yield. This result was rather surprising because we expected a 5-*exo* cyclization to be faster than any second supposedly intermolecular trapping of the radical by vanadium. Because of the dimeric nature of the vanadium reagent in the solid state, the second electron could be transferred to the radical center in an intramolecular manner. This process should be able to compete efficiently with the radical cyclization [40, 41]. This interpretation is in line with the 50 : 50 mixture of olefins obtained in the deoxygenation of **7** with vanadium via a radical intermediate (Table 1, entry 2). A rigorous confirmation of this hypothesis could be obtained from solution studies of the reagent's structure. It should be noted, however, that preliminary kinetic investigations of the reaction of $[V_2Cl_3(THF)_6]_2[Zn_2Cl_6]$ with benzaldehyde indicated that the reactive species in solution is VCl_2 (Wulff and Daasbjerg, unpublished results).

Entries 4 and 5 in Table 2 readdress the issue of the intramolecular nucleophilic epoxide opening leading to the formation of **15**. The precursor to the vanadium(II) reagent, $VCl_3(THF)_3$, and VCl_3/Zn in THF result in the formation of **15** but in higher yields than with $[V_2Cl_3(THF)_6]_2[Zn_2Cl_6]$. Since both reagents constitute stronger Lewis acids than $[V_2Cl_3(THF)_6]_2[Zn_2Cl_6]$, vanadium(III) species are likely to be responsible for the undesired side reaction and not the vanadium(II) complex. Titanocene chloride gave **16** in 57% yield. This result can be readily explained by an epoxide opening to yield radical **11** ($M = Cp_2TiCl$) via ET and the ensuing 5-*exo* cyclization. The second ring is closed through a radical substitution reaction that will be discussed later.

To confirm the trends observed with **10**, we also investigated the behavior of epoxide **17** under ET conditions. Here, a tertiary radical would be formed after reductive opening that is more persistent than the secondary radical obtained from **10**, as depicted in Scheme 8. The results of the opening reactions are summarized in Table 3.

It is well documented that tertiary radicals are only slowly reduced even by potent electron-transfer reagents, e.g., SmI_2 [42, 43]. Therefore, we expected **17** to be a better substrate for radical cyclization than **10**. However, neither $[V_2Cl_3(THF)_6]_2[Zn_2Cl_6]$ nor SmI_2 gave any of the desired product. Utilizing the vanadium-based reagent, only the product of deoxygenation **18** could be obtained in significant amounts. Trace amounts of the chlorohydrins were detected in the crude reaction mixture. Thus, $[V_2Cl_3(THF)_6]_2[Zn_2Cl_6]$ constitutes a highly selective reagent for the deoxygenation of epoxides, and does not allow other synthetic applications of the pivotal β-metal oxy radical formed during reductive epoxide opening. SmI_2 gave the primary iodohydrin **19** in 45% yield and deoxygenation product **18** in 36% yield. As in the

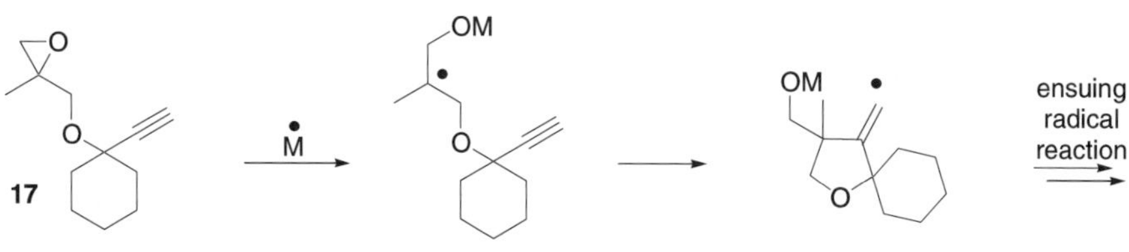

Scheme 7 Radical reactions of **17**

Table 3 Competing pathways in the opening of **17**

Entry	ET reagent	Product	Yield (%)
1	$VCl_3(THF)_3/Zn$	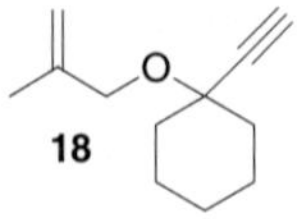 **18**	57
2	SmI_2	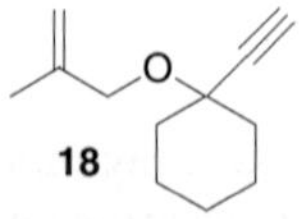 **18**	36
		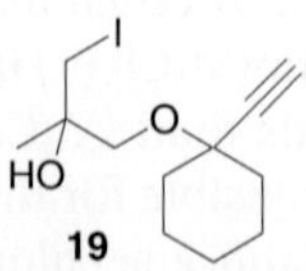 **19**	45
3	Cp_2TiCl	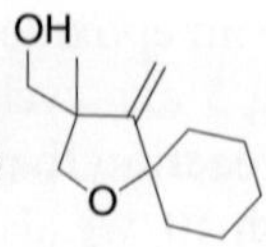 **20**	69-82
4	$CrCl_2$	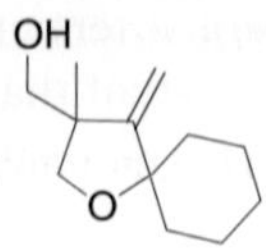 **20**	38[a]

[a] 37% substrate reisolated

reactions of **10**, nucleophilic opening of the epoxide is the main course of events. The samarium salt of iodohydrin **19** can be further reduced by SmI_2 to afford **18**.

The result obtained with $CrCl_2$ was rather surprising. As in the deoxygenation reaction the substrate **17** could be reisolated in substantial amounts. However, the product of the radical cyclization **20** was also obtained in noticeable amounts (38%). Thus, $CrCl_2$, although it is rather unreactive, is the only reagent other than Cp_2TiCl that results in the formation of the β-metal oxy radical through direct ET, which can be utilized in a subsequent C–C bond-forming reaction. Our titanocene-based protocol, on the other hand, gave the desired cyclization product **20** in good yield (69–82%), and once again demonstrates the superiority of the titanocene(III) reagents in reductive epoxide openings.

In summary, our studies have revealed that SmI_2 is not a suitable reagent for the reductive opening of epoxides. The high Lewis acidity of this metal combined with the high nucleophilicity of the iodide ions leads to the formation of iodohydrins that are further reduced by samarium. The typical reactivity of β-metal oxy radicals as shown in Schemes 7 and 8 could not be observed in the cases investigated here. With $[V_2Cl_3(THF)_6]_2[Zn_2Cl_6]$, no products other than those of epoxide deoxygenation could be observed. Although it seems as if this reagent opens epoxides via ET, the resulting β-metal oxy radicals could not be intercepted by C – C bond-forming reactions. This unexpected result indicates that the radical is further reduced in a fast second ET reaction, which might be tentatively rationalized by assuming a dimeric structure of the vanadium reagent in solution thereby allowing an intramolecular second electron transfer. Our results strongly suggest that the reason for the superiority of Cp_2TiCl reagents stems from their unique combination of low Lewis acidity preventing epoxide opening via S_N2 or S_N1 reaction and low reducing power toward the β-metal oxy radical. The only other reagent that allows C – C bond formation, $CrCl_2$, is unfortunately severely limited by its low reactivity, most likely caused by its low solubility. Ligand variation should lead to more reactive complexes, however.

3 Mechanism of Titanocene-Mediated Epoxide Opening by Homolytic Substitution: Catalyst Structure, Kinetics and Thermodynamics, Synthetic Implications

All experimental results are in line with formation of the pivotal β-titanoxy radicals through a homolytic substitution reaction of epoxides and titanocene(III) complexes. This finding is of great practical and mechanistic relevance, as the stage is now set for combining the advantages of Lewis acid catalysis and radical chemistry. This is especially so for the design of highly selective titanocene reagents. However, hardly anything is known about catalyst structure in solution, the thermodynamic and kinetic features of epoxide opening, and the factors affecting regioselectivity. This understanding is essential from a fundamental point of view, for catalyst improvement and for novel synthetic applications. We investigated these mechanistic questions by a combination of cyclic voltammetry, kinetic studies, computational chemistry, and preparative means.

3.1 Cyclic Voltammetry

The determination of catalyst composition and the kinetics of epoxide opening constitute the experimental basis for any mechanistic discussion and catalyst design. For this purpose, Zn-reduced THF solutions of the preparatively important **21** [44–46], **22** [47], **23**, and **24** were analyzed by cyclic voltammetry, a technique uniquely suited to the investigation of redox active

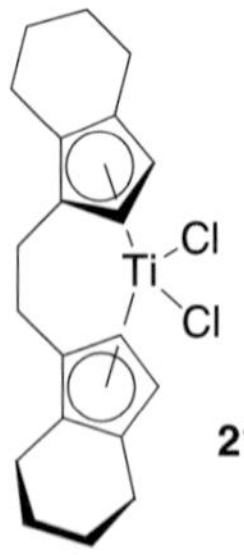

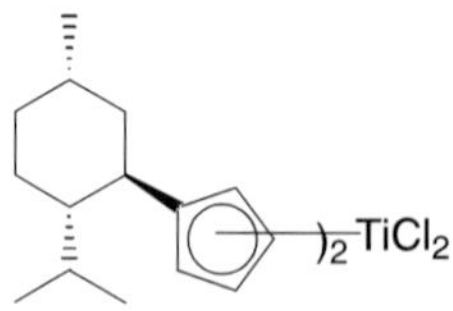

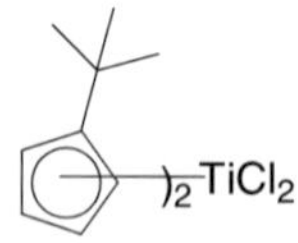

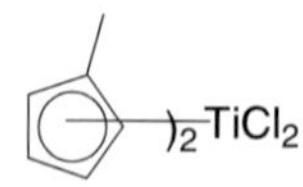

21 **22** **23** **24**

species in solution [48]. The kinetics of epoxide opening was studied by UV spectroscopy.

Previously, it has been shown that the Cp_2TiCl_2-derived solution consists of two species, Cp_2TiCl and the chlorine-bridged dimer $(Cp_2TiCl)_2$, which are in rapid equilibrium [49–51]. This was established by carrying out a detailed cyclic voltammetric analysis at different concentrations, as shown in Fig. 2a. The broad oxidation wave appearing at -0.8 V vs Fc^+/Fc (ferrocenium/ferrocene) actually consists of two processes, with the first one becoming more dominant as the concentration increases. Thus, the first peak is attributed to the oxidation of the $(Cp_2TiCl)_2$ dimer and the second peak to the oxidation of the Cp_2TiCl monomer. The oxidation peak at -0.4 V vs Fc^+/Fc and the reduction peak appearing at -1.2 V vs Fc^+/Fc on the reverse sweep are due to the generation of Cp_2Ti^+ and Cp_2TiCl_2, respectively, at the electrode surface during the cyclic voltammetric sweep. In other words, Cp_2Ti^+, Cp_2TiCl_2, trinuclear species containing two Ti and one Zn atom, or ionic clusters, are not present in solution.

On the basis of such an analysis the dimerization constant of Cp_2TiCl was determined to be 3×10^3 M^{-1}. This implies that for Ti(III) concentrations above 1 mM the dimer becomes the dominant species. However, the Zn-reduced solutions of the other and more sterically hindered complexes **21**–**24** are monomeric, the exception being that of **24**, since the appearance of the voltammogram remains essentially the same on changing the concentration. In Fig. 2b this is illustrated for the case of **21**.

In general, the Zn-reduced solutions of the higher substituted complexes are better ET reagents thermodynamically, as indicated by the standard potentials of the oxidized forms listed in Table 4. The nature of the reacting complexes was established by measuring the rate constants of the opening of **25** with the Zn-reduced solutions of all titanocenes denoted "Ti" and the reduction of the radicals carried out with 1,4-cyclohexadiene, as outlined in Scheme 8. This was achieved by monitoring the disappearance of the Ti(III) species with a UV dip probe using an excess of the epoxide.

The reaction rates that are also summarized in Table 4 are independent of the presence of 1,4-cyclohexadiene. Epoxide opening therefore constitutes the rate-controlling step of the overall reduction. This renders any mechan-

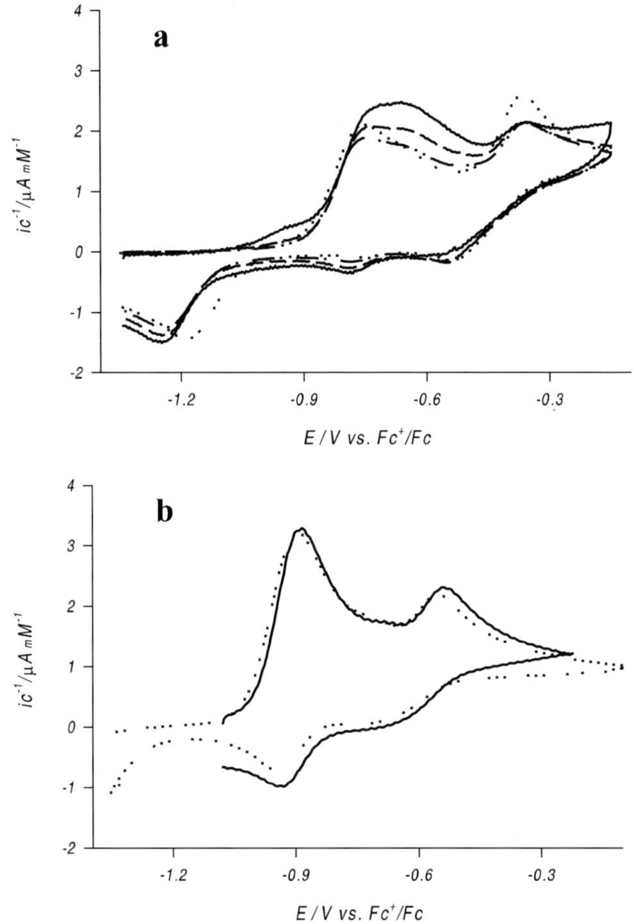

Fig. 2 Cyclic voltammograms recorded at a glassy carbon disk electrode (diameter = 1 mm) at a sweep rate of 1 V s^{-1} in 0.2 M Bu_4NPF_6/THF for Zn-reduced solutions of **a** Cp_2TiCl_2 at concentrations of 0.2 (—), 0.4 (– – –), 1 (–··–), and 2 mM (· · ·) and **b 21** at concentrations of 0.7 (—) and 1.5 mM (· · ·)

ism with a quick and reversible epoxide opening before radical trapping by the hydrogen atom donor highly unlikely. Within experimental error the rate constants are independent of the concentrations of **21**, **22**, and **23**. Thus, in these cases the monomers are reactive species and no dimers are involved at all. Except for **23**, where substrate binding seems to be sterically hindered by the *tert*-butyl groups, the thermodynamically better reductants also open the epoxides more swiftly.

In the case of Zn-reduced Cp_2TiCl_2, the relative reaction rate increases with increasing concentration of the titanium species. Thus, it may be concluded that both the monomer and the dimer are able to open the epoxide.

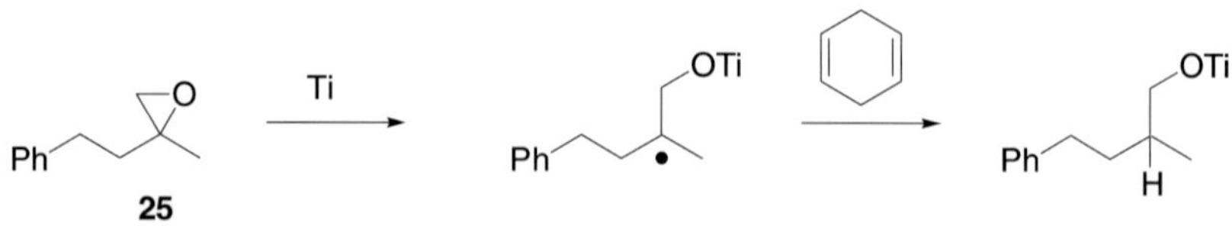

25

Scheme 8 Reductive epoxide opening employed in the kinetic studies

Table 4 Standard potentials E^o extracted from cyclic voltammograms recorded for the dimeric and monomeric species present in Zn-reduced solutions of Cp_2TiCl_2 and **21**–**24** along with rate constants k measured for their reduction of **25** in THF

	$(Cp_2TiCl)_2$/ Cp_2TiCl	**21** monomer	**22** monomer	**23** monomer	**24** dimer/ monomer
E^o (V vs Fc^+/Fc)	−0.81/−0.75	−0.94	−0.84	−0.81	−0.91/−0.81
k ($M^{-1}\,s^{-1}$)	1.4/0.5	4.8	0.8	0.1	3.9/1.3

By numerically fitting the decay curves of [Ti(III)] with the simulation program Gepasi [52], it was established that the dimer opens the epoxide with a rate constant of $k = 1.4\,M^{-1}\,s^{-1}$, whereas the monomer reacts more slowly ($k = 0.5\,M^{-1}\,s^{-1}$). At the usual initial Cp_2TiCl_2 concentration of 10 mM, this means that 84% of **25** molecules are opened by the dimer.

At first glance such a result may seem surprising, as the dimer contains no vacant binding site. However, in the formation of $(Cp_2TiCl)_2$ [49–51, 53, 54] from Cp_2TiCl, the half-open structure **26**, possibly solvated by THF as in **27**, constitutes an intermediate as shown in Scheme 9. Both **26** and **27** are substantially stronger Lewis acids than $(Cp_2TiCl)_2$, Cp_2TiCl, and Cp_2TiCl^*THF according to the principle of activation of electrophiles by electrophiles through dimeric association [55]. This will result in a faster formation of epoxide–titanocene complexes, and possibly also affect the activation energies of epoxide opening. This point will be discussed later with the aid of computational chemistry.

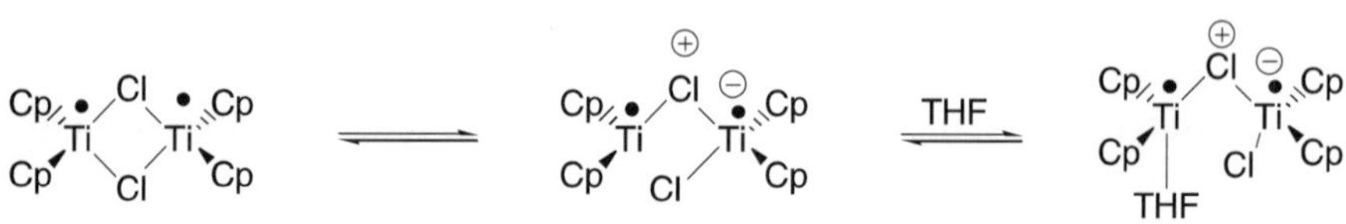

26 **27**

Scheme 9 Half-open dimers involved in epoxide opening by Zn-reduced Cp_2TiCl_2

An even higher reactivity difference of the dimer and monomer has already been observed in pinacol couplings of benzaldehyde using Zn-reduced solutions of Cp_2TiCl_2 ($k_{dimer} = 70\,M^{-1}\,s^{-1}$, $k_{monomer} < 2\,M^{-1}\,s^{-1}$) [50]. The higher reaction rates observed for pinacol coupling compared with epoxide opening are explained by a faster coordination of the less hindered and more Lewis basic aldehyde by the titanocene species involved. Therefore, even at a 0.5 mM concentration of Cp_2TiCl_2, 85% of the pinacol coupling proceeds through the dimer.

In summary, the voltammetric and kinetic studies have delivered the essential knowledge about catalyst composition and rate constants of epoxide opening necessary for the computational studies. Moreover, a reversible epoxide opening has been rendered unlikely.

3.2 Computational Investigations and Catalyst Design

With the aid of these experimental findings it is possible to conduct meaningful computational investigations. We studied the reaction and activation energy of epoxide opening, and the structures of the relevant intermediates and transition states, by density functional theory (DFT) calculations with the B-P functional and a TZVP basis set [56–58],[1] and started our investigations with Cp_2TiCl complexed with ethylene oxide (**28**). This complex and its reactions will also serve as structural models for the substituted titanocenes that are monomeric in solution. As shown in Scheme 10, replacement to give **29** is essentially thermoneutral. In comparison, formation of **30** is disfavored by about 4 kcal mol^{-1}. This constitutes a clear indication of the strong steric interactions of the epoxides within the binding site of Cp_2TiCl. The results are shown in Scheme 10.

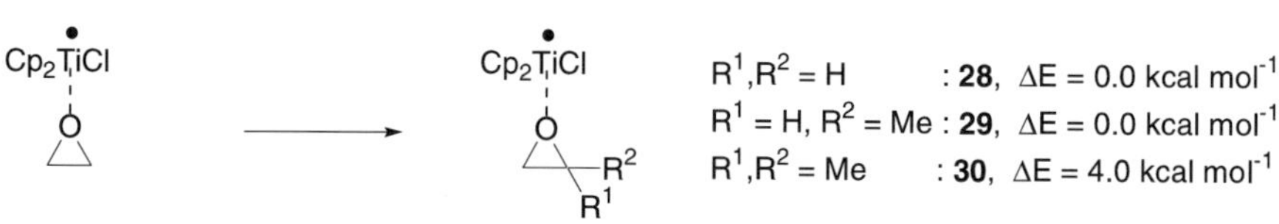

Scheme 10 Reaction energies of epoxide binding by Cp_2TiCl

According to the kinetic CV measurements, epoxide opening through ET constitutes the rate-controlling step of reductive epoxide opening. The reac-

[1] All quantum chemical calculations were performed with the TURBOMOLE suite of programs. The structures were fully optimized at the DFT level by employing the BP86 functional, a Gaussian AO basis of triple-zeta valence quality including polarization functions (TZVP) and the RI approximation for the two-electron integrals. For titanium an all-electron basis set of triple-zeta valence quality ([6*s*4*p*3*d*]) was employed. The nature of the transition states was checked by calculation of the harmonic vibrational frequencies, and in all cases one imaginary vibrational mode corresponding to the expected reaction coordinate was found.

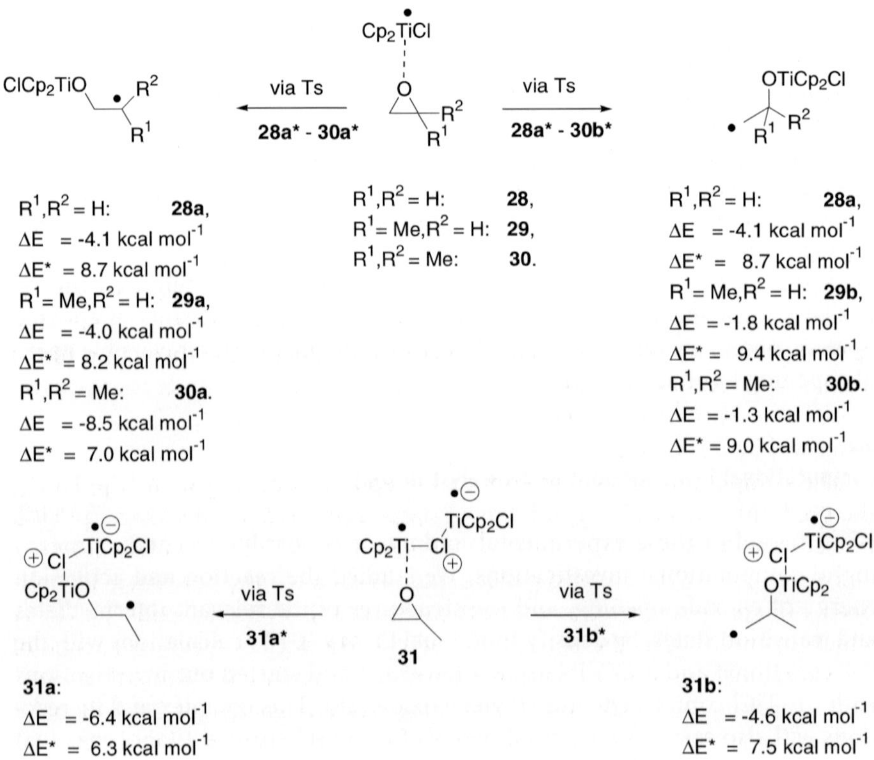

Scheme 11 Reaction and activation energies of epoxide opening by DFT methods

tion and activation energies of the step are essential for an understanding of the reaction, and therefore we calculated their values. The results are summarized in Scheme 11. The values of $-\Delta E$ for ring opening of the titanocene epoxide complexes are in the range of 1–9 kcal mol^{-1}. This is unexpectedly low when considering the high strain of oxiranes (ca. 25 kcal mol^{-1}) and the formation of the supposedly strong Ti–O bonds. It seems that these two energetically favorable contributions are just enough to compensate the stability difference between the titanium-centered and carbon-centered radicals.

Quite surprisingly, **28a** and **29a** are formed from **28** and **29** with about the same reaction energy ($\Delta E \approx -4.0$ kcal mol^{-1}), even though secondary radicals are more stable than primary radicals by approximately 3 kcal mol^{-1} based on their bond dissociation energies. This must be due to steric interactions with the cyclopentadienyl ligand in **29a**, which fully counterbalances the radical's increased stability. A similar trend of product stability is observed in the formation of the less favored primary radicals **29b** and **30b**. The formation of **30a** is more favorable by 4.5 kcal mol^{-1} compared to **29a**. This is even higher than the stability difference between a tertiary and a secondary radical. Generation of the looser structure of **30a** releases some of the energy associated with the less favorable complexation.

Relative to the epoxide complexes, all activation energies for ring opening are in the range of 7.0–9.5 kcal mol^{-1} indicating a fast radical generation at room temperature. In the transition structures the spin density on the evolving radical center is only about 0.2, and about 0.8 on titanium. Therefore, the factors governing the stability of carbon-centered radicals should not dominate the relative stabilities of the transition states. As main reasons for these energy differences favoring **29a** and **30a** at the expense of **29b** and **30b**, respectively, we suggest unfavorable steric interactions between the epoxides' methyl groups and the cyclopentadienyl ligands during the formation of the less substituted radicals. If this notion is correct, substituted titanocene complexes should lead to an epoxide opening with higher regioselectivity. This point will be discussed later.

The computational studies concerning epoxide opening with $(Cp_2TiCl)_2$, the predominant species in solution that opens epoxides faster than Cp_2TiCl, turned out to be revealing, too. We were unable to detect a stable complex between $(Cp_2TiCl)_2$ and propene oxide. Therefore, the reactions of half-open dimers were investigated next (Scheme 11). Propene oxide yields **31** which opens to give the secondary radical **31a** with a reaction energy of −6.4 kcal mol^{-1} and an activation energy of 6.3 kcal mol^{-1}. For **31b** a reaction energy of −4.3 kcal mol^{-1} and an activation energy of 7.5 kcal mol^{-1} were calculated. Both activation energies are noticeably lower than that for Cp_2TiCl, which is in agreement with the measured higher rate constants for $(Cp_2TiCl)_2$. However, the differences in reaction and activation energies, which are responsible for the observed selectivities after all, between the respective species in the case of **29** and **31** are essentially the same. This indicates that the same factors controlling the selectivity are operating in both cases. This finding supports our steric model, as all structures are very similar in the near surroundings of the complexed propene oxide. The second Cp_2TiCl unit in **31** is pointing away from the reaction center. Thus, the simpler structures involving Cp_2TiCl can be used as good models for devising a more selective catalyst.

The calculated reaction and activation energies can also be used to distinguish between an irreversible and a reversible epoxide opening. Based on the reaction energies, ratios of 97 : 3 for **29a** : **29b**, 96 : 4 for **31a** : **31b**, and > 99 : < 1 for **30a** : **30b** are predicted at 25 °C for a thermodynamically controlled opening. For a kinetically controlled opening the activation energy differences are decisive for regioselectivity. At 25 °C a ratio of 88 : 12 for both **29a** : **29b** and **31a** : **31b**, and of 97 : 3 for **30a** : **30b**, is predicted. Again, the simpler Cp_2TiCl-based system can be used for predictions without substantial error.

Experimentally, the reactions of 1,1-disubstituted epoxides were not studied to address these issues because the predicted differences in selectivity are too small to be verified. Instead, the reactions of **32**, a nonvolatile substitute of propene oxide, were carried out as shown in Scheme 12. In this case larger differences are expected.

Epoxide opening proceeded with a regioselectivity of 88 : 12 to 90 : 10 for both **33** and **34**. These values are in excellent agreement with the calculated

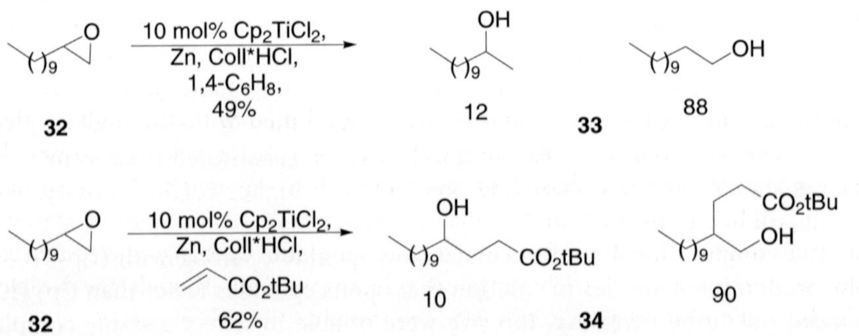

Scheme 12 Opening of 32 by the Zn-reduced solution of Cp_2TiCl_2

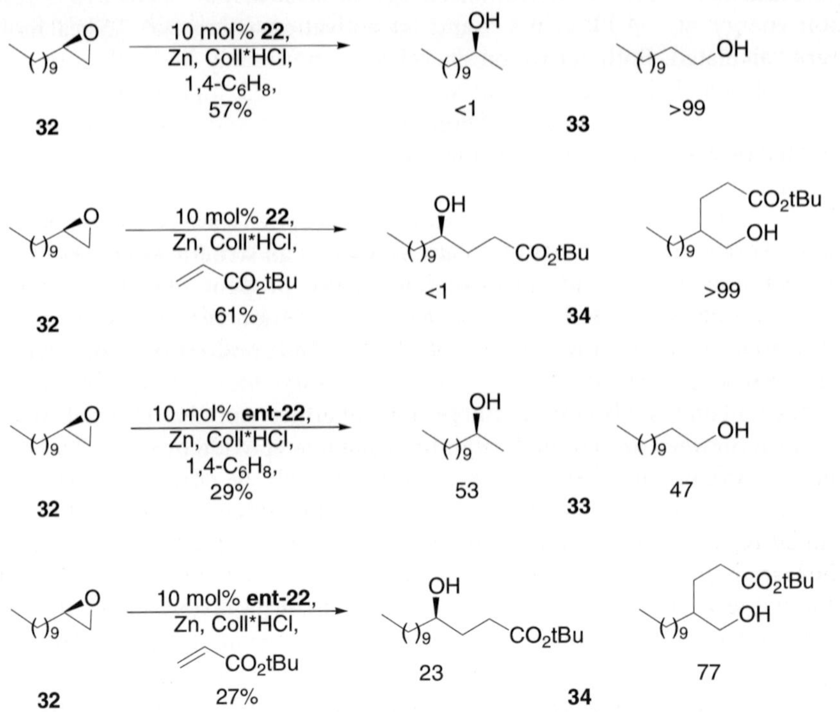

Scheme 13 Matched and mismatched cases of regioselectivity in the opening of 32 by 22

values of an irreversible epoxide opening (88 : 12) and are far too low for a reversible reaction, for which selectivities of 97 : 3 were predicted. Thus, the DFT calculations constitute an excellent tool for studying the mechanism of the titanocene-catalyzed or -mediated epoxide opening.

Since we have suggested steric interactions between the Cp ligands and the epoxide as the reason for the regioselectivity of epoxide opening, we also studied **22** in the opening of **32** in order to verify this assumption. The results are summarized in Scheme 13. A matched and mismatched case of both regioselectivity and yield of the desired products was observed for the reaction of enantiomerically pure **32** with both enantiomers of **22**. In the former case very high regioselectivity combined with the highest yield of the desired products was observed. The remainder was constituted by **32**, and thus the yield based on recovered starting material was > 90%, which excludes the fact that undesired side reactions occurred to a substantial degree. In the mismatched case epoxide opening was distinctly slower and proceeded with very low selectivity. Even though the isolated yields of the desired products were low, the yield based on recovered starting material was also > 90%. Thus, it seems that the steric interactions that prevent quick substrate binding also lead to an unselective epoxide opening. These findings will be of substantial practical importance for the development of kinetic resolution reactions and for the application of reductive epoxide opening in the synthesis of enantiomerically pure natural products.

3.3 Enantioselective Opening of *meso*-Epoxides

Our hypothesis of steric factors dominating the stability of the emerging radical centers in the transition states readily explains the enantioselective epoxide opening of *meso*-epoxide **35** to **36** that is shown in Fig. 3 [59, 60]. In the case of a reversible epoxide opening, a stability difference of at least 3 kcal mol^{-1} between the two radicals **37** and **38** is necessary to explain the observed selectivity. According to the calculations this seems highly unlikely. A thermodynamically controlled epoxide opening can therefore be ruled out.

The assumption of a kinetically controlled course of the reaction, however, readily explains the observed results, even though the transition structures have not, as yet, been calculated. Because epoxide opening is exothermic, **39** can be regarded as a simple model for the transition structure according to the Hammond postulate. It is clear from the structure of **39** that the left-hand ethoxy substituent of the epoxide is in close proximity to the ligand of the catalyst, whereas the other substituent hardly encounters any steric interaction. Epoxide opening will release the former interaction. After reduction of the radical, this results in formation of the product with the absolute configuration observed experimentally.

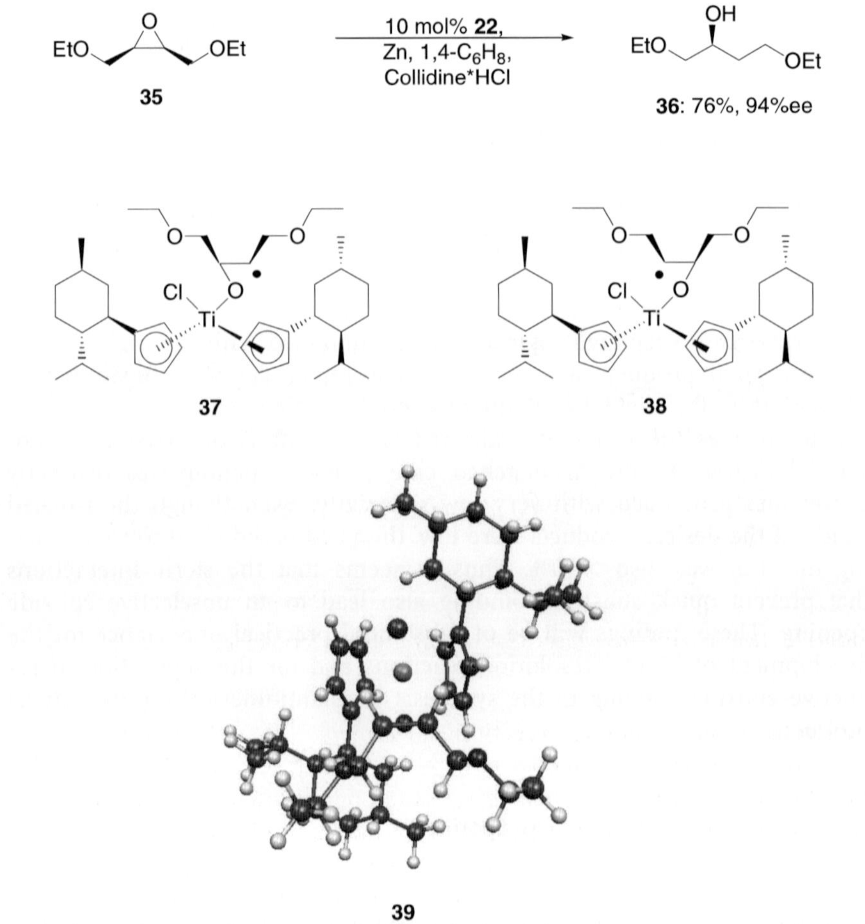

Fig. 3 Enantioselective opening of *meso*-epoxide 35

4 Formation of Tetrahydrofurans

The surprisingly low exothermicity of epoxide opening, which is most likely due to the high stability of the titanium-centered radical, suggests that the closure of relatively unstrained rings might be possible [61, 62]. As an example, this is shown for tetrahydrofuran formation in Scheme 14. Such a transformation is complementary to the ET-induced ring opening of tetrahydrofurans under Lewis acid catalysis by arene radical anions, as introduced by Mudryk and Cohen [63].

Of course, an efficient method for the generation of the necessary radicals is essential for a useful application of this reaction. We decided to

Scheme 14 Tetrahydrofuran formation via S_H2 mechanism

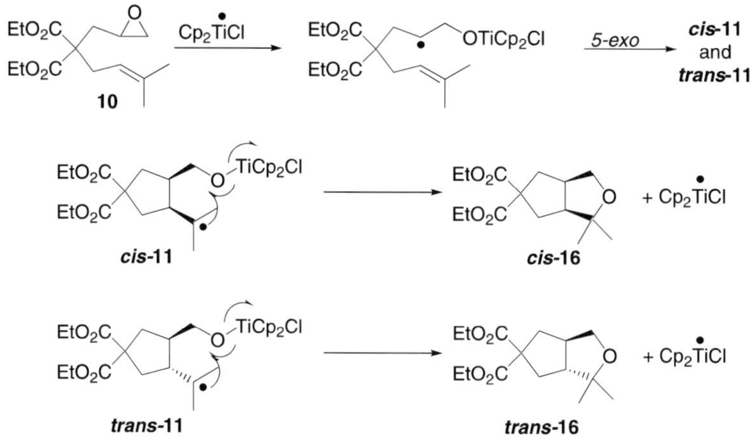

Scheme 15 Radical tandem sequence for tetrahydrofuran formation

implement a radical tandem reaction (for a review, see [64]) that meets this demand by the sequence shown in Scheme 15. The sequence is initiated by the titanocene-mediated reductive epoxide opening [65, 66] of **10** followed by a 5-*exo* cyclization onto a trisubstituted olefin to obtain the crucial radical **11**. In this intermediate the radical center is positioned appropriately for ring closure to yield the desired tetrahydrofuran.

Besides the thermodynamic feasibility of the Ti – O bond rupture, the possibility of the formation of the two possible bicyclic tetrahydrofurans deserves further consideration. In the case of *trans*-**16** a strained and thus probably unstable bicyclo[3.3.0] system would be formed.

4.1 Computational Investigations

In order to clarify these points and to verify the practicability of our concept, we investigated the course of the overall transformation by density functional theory (DFT) as depicted in Scheme 16. As expected, both the ring opening of **40** to **41** and the 5-*exo* cyclizations to **42** are exothermic. The calculated Ti – O bond lengths (1.86 Å) are in excellent agreement with values obtained from crystallographic structures (1.85–1.89 Å) [67–69].

The calculations suggest that the well-documented kinetically controlled course of 5-*exo* cyclizations that has been manifested in the Beckwith–Houk

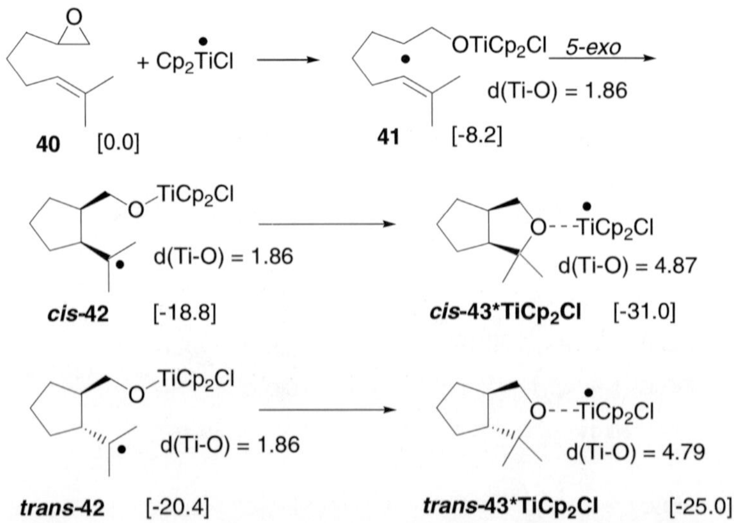

Scheme 16 Reaction energies of tetrahydrofuran formation via S_H2 mechanism

rules [39, 70] is occurring in our case as well. Thus, despite the thermodynamic preference for the formation of *trans*-**42** (– 12.3 kcal mol^{-1}), the other diastereoisomer *cis*-**42** (– 10.6 kcal mol^{-1}) is formed preferentially due to an energy of activation (6.1 kcal mol^{-1} for *cis*-**42** and 7.6 kcal mol^{-1} for *trans*-**42** relative to **41**) that is lower by 1.5 kcal mol^{-1} according to our calculations.

The tetrahydrofuran-forming reactions that yield the titanocene(III) complexes of *cis*- and *trans*-**43** were analyzed next. Both reactions were calculated to be exothermic (– 12.2 and – 4.6 kcal mol^{-1}, respectively). As a plausible reason for this finding we suggest the high stability of the titanium-centered radical Cp_2TiCl. The less negative reaction energy found in the case of *trans*-**43** reflects the strain of the *trans*-fused bicyclo[3.3.0] system.

The geometries of the optimized product complexes **43***Cp_2TiCl, which are only slightly more stable than the separated molecules, show a very loose coordination between the tetrahydrofuran and titanium. The accurate determination of this complexation energy is beyond the scope of the applied DFT methods, and the numbers given are also influenced by basis set incompleteness effects. This is not, however, of much practical relevance, since entropy effects will favor the dissociation of the product complex and thus the regeneration of Cp_2TiCl. Of course, this weak coordination is beneficial for catalysis because ligand exchange with the solvent or the substrate is easy to achieve.

We have also calculated the transition state for the ring closure of *cis*-**42** to *cis*-**43***$TiCp_2Cl$, which is shown in Fig. 4. The final ring closure has a relatively low barrier (+ 11.4 kcal mol^{-1}) and should therefore be liable even at low temperatures. The transition state exhibits similar Ti – O and C – O bond lengths. This indicates that a homolytic, concerted substitution reaction (S_H2) resem-

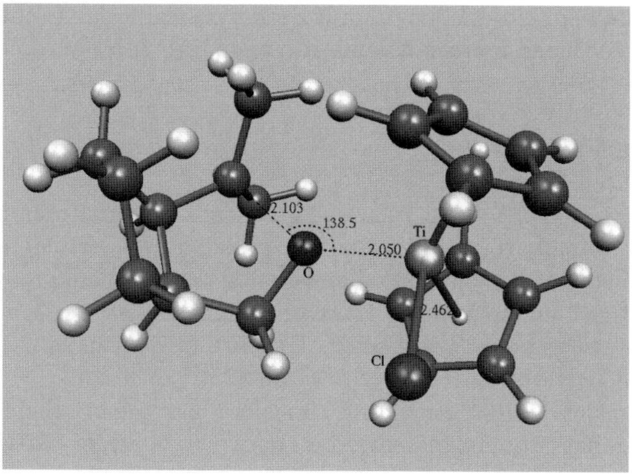

Fig. 4 Transition structure in the formation of *cis*-**43***$TiCp_2Cl$

bling the S_N2 reaction is taking place. However, the criteria for back-side attack seem to be less stringent for the S_H2 mechanism because the C – O – Ti angle amounts to about 140° as compared to the 180° necessary for nucleophilic substitution (for a discussion of the mechanism of the S_N2 reaction, see [71]). Also the structure is fairly compact and one can expect the reaction to be sensitive to steric effects.

Qualitatively, the transition structure of the cyclization would be expected to be early according to the Hammond postulate [72]. However, matters are more complicated. The electronic structure of the transition state is revealing in this context. According to the spin densities from a Mulliken population analysis, the radical center is already shifted from carbon to titanium (C +0.37, O −0.05, Ti +0.70) even though the developing C – O bond is very long (2.10 Å vs about 1.43 Å for Csp^3 – O). The Ti – O bond (2.05 Å) is still mostly intact. The transition structure must therefore be considered as "electronically late", although according to the bond lengths it must be regarded as early. Thus, the notion that the stability of the titanium-centered radical constitutes a major driving force for the cleavage of the Ti – O bond is already apparent in the transition structure of tetrahydrofuran formation.

The transition structure *trans*-**42** to *trans*-**43***$TiCp_2Cl$ was also calculated and a barrier of 11.7 kcal mol^{-1} was obtained for the cyclization. This value is similar to the activation energy for the formation of *cis*-**43***$TiCp_2Cl$, although the distance between the carbon-centered radical and oxygen atom is shorter (2.04 Å).

In summary, according to the computational analysis it should be possible to realize the desired overall tandem sequence in the absence of alternative radical pathways with lower energies of activation. The formation of the *cis*-substituted bicyclo[3.3.0] system is both kinetically, due to the faster formation of *cis*-**42**, and thermodynamically more favorable.

4.2 Synthetic Implications

We have chosen epoxy olefin **10** as substrate for our initial examinations for two reasons. Firstly, **10** is synthesized in a straightforward manner from allyl diethyl malonate by epoxidation and an S_N2 reaction with prenyl bromide. Secondly, it is known from the work of others [17–20] and ourselves [65, 66, 73, 74] that compounds similar to **10** cyclize to yield mainly the essential *cis*-fused radicals with selectivities of about 85 : 15 to 90 : 10.

It is also essential that competing radical pathways are excluded. The radical intermediates should therefore be relatively persistent. This is the case here, because tertiary radicals are relatively slowly trapped by hydrogen atom donors, e.g., THF, which is usually applied as solvent in titanocene-mediated or -catalyzed reactions, or a second equivalent of Cp_2TiCl. However, in the absence of other pathways this reduction, which was followed by a β-hydride elimination, was observed [75, 76]. Our results with **10** are summarized in Table 5.

The catalytic reaction conditions required some optimization. This was due to competing reaction pathways. The interception of *trans*-**11** results in the formation of the organotitanium intermediate **44**, as shown in Scheme 17. Thus, 2 equiv. of Cp_2TiCl are consumed and a complete conversion in the presence of 10 mol % Cp_2TiCl_2 cannot be achieved because catalyst regeneration is prevented. Similar considerations apply for *cis*-**11**.

The organometallic compound **44** should decompose via β-hydride elimination to olefin **45a** and hydrido titanium species **46** in the absence of acid. The catalytic cycle is interrupted due to the consumption of Cp_2TiCl. In the presence of Coll*HCl, protonation of the Ti – H and Ti – O bonds in **46** and

Table 5 Optimization of the titanocene-catalyzed transformation of **10** to *cis*-**16** at room temperature. All reactions are carried out at 0.1 M concentration

Entry	Catalyst loading (% mol)	Conditions	Yield (%)
1	10	THF, Zn (2.0 eq), Coll*HCl (2.5 eq)	57
2	10	THF, Mn (0.2 eq), Coll*HCl (0.5 eq)	67
3	10	THF, Mn (0.2 eq), Coll*HCl (0.5 eq)	67 [a]
4	10	THF, Zn (0.2 eq), Coll*HCl (0.5 eq)	52
5	10	EtOAc, Zn (2.0 eq), Coll*HCl (2.5 eq)	62

[a] 4 h, 70 °C

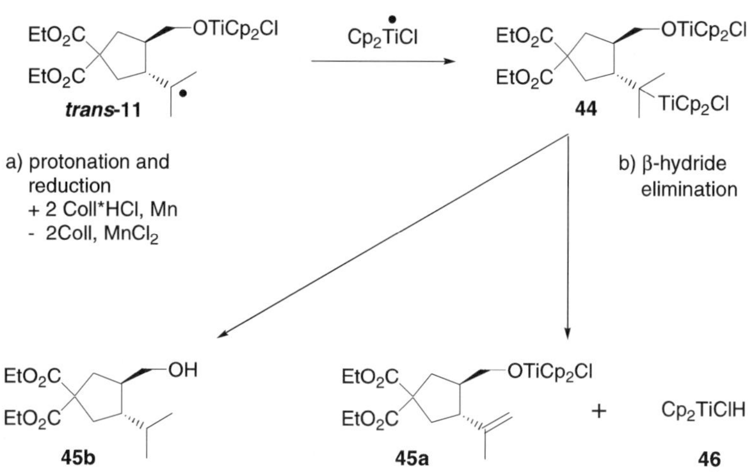

Scheme 17 Possible side reactions of tetrahydrofuran formation

45a liberates Cp_2TiCl_2. Regeneration of Cp_2TiCl is accomplished by the metal powder employed as reductant. Additionally, a direct interception of **44** by double protonation to yield **45b** is also possible. In this case the Cp_2TiCl_2 formed also needs to be reduced to regenerate the redox catalyst. Thus, 2 equiv. of Coll*HCl and 1 equiv. of the reductant are needed to regenerate Cp_2TiCl from this competing side reaction. Similar considerations apply for *cis*-**11** in case the tetrahydrofuran formation is relatively slow. In all cases examined mixtures of *cis*- and *trans*-**45a** and **45b** were obtained as side products, which accounted for 30–35% of the consumed starting material. The exact ratios of the isomers could not be determined.

When the reaction was run under the typical conditions of reductive epoxide opening [65, 66, 73, 74] a yield of 57% was obtained (Table 5, entry 1). As expected, 2 equiv. of the reductant and 2.5 equiv. of Coll*HCl are not needed for complete consumption of the starting material. With 0.2 equiv. of metal powder and 0.5 equiv. of the acid, the best results were obtained with Mn in THF (entry 2). Heating of the reaction mixture resulted in a faster transformation without reduction in yield (entry 3). In the case of Zn as reductant, EtOAc turned out to be superior to THF as solvent (entries 4 and 5).

We suggest that the superiority of Mn to Zn and of EtOAc to THF in the case of Zn is due to a slower reduction of Cp_2TiCl_2 to Cp_2TiCl (Zn: ca. 1 min in THF, 30–60 min in EtOAc; Mn: 3–5 min in THF, > 60 min in EtOAc) which can be observed by a color change from red to green. A lower concentration of the reductant will decrease the rate of the undesired bimolecular interception of *cis*-**11** (and *trans*-**11**) and therefore increase the yield of *cis*-**16** (Scheme 4).

We also investigated three other complexes as potential catalysts for the preparation of *cis*-**16** from **10**. The results are summarized in Table 6. While **24**

Table 6 Comparison of different catalysts in the preparation of **16**

Entry	Catalyst	Conditions	Yield (%)
1	Cp_2TiCl_2	THF, Mn (2.0 eq), Coll*HCl (2.5 eq)	55
2	$(MeCp)_2TiCl_2$ (**24**)	THF, Mn (2.0 eq), Coll*HCl (2.5 eq)	46
3	$(t\text{BuCp})_2TiCl_2$ (**23**)	THF, Mn (2.0 eq), Coll*HCl (2.5 eq)	20

performed reasonably well (entry 2), the bulky **23** gave a disappointing yield (entry 3). The compact transition structure (Fig. 4) supports the assumption that the system should be sensitive toward steric interactions.

4.3 Scope and Limitations of Tetrahydrofuran Formation

Our method is attractive from a synthetic point of view as the molecular complexity is substantially increased from simple starting materials in a single step [75, 76]. Therefore, to analyze the mechanistic implications, we decided to investigate the influence of the olefin substitution pattern on the outcome of the reaction.

As summarized in Table 7, a number of trisubstituted olefins can act as suitable radical precursors if the substituents are not too bulky. The synthesis of spirotricyclic compounds, e.g., **48**, is readily achieved in yields comparable to that of **16** (entry 1). To the best of our knowledge no simple method for the synthesis of these complex products is available. The bicyclo[4.3.0] system

Table 7 Trisubstituted olefins as substrates in tetrahydofuran formation

Entry	Substrate	Product	Yield (%)
1	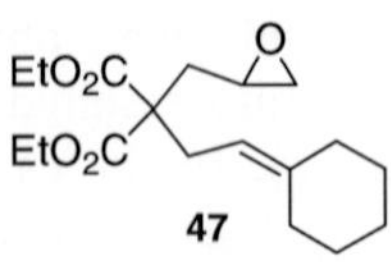 **47**	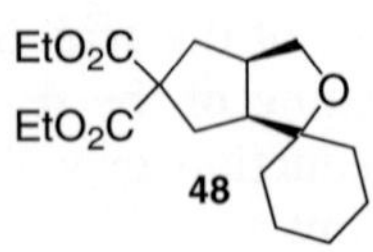 **48**	62
2	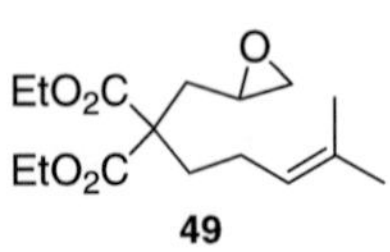 **49**	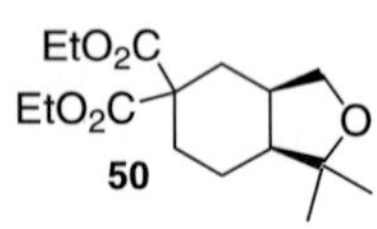 **50**	62, dr = 87:13
3	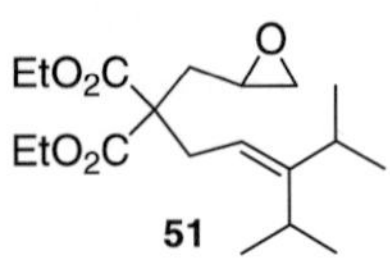 **51**	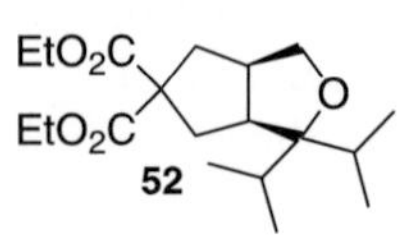 **52**	0

50 (entry 2) demonstrates the specific usefulness of our titanocene-catalyzed protocol. Usually 6-*exo* cyclizations are distinctly less efficient than the corresponding 5-*exo* cyclizations, in contrast to our system [77, 78]. In the case of the bis-isopropyl-substituted olefin **51** (entry 3), none of the desired product **52** was obtained. The same holds true for *tert*-butyl-substituted olefins (not shown). The approach of the intermediate radicals toward the Ti – O bond is too hindered to enable an S_H2 reaction.

The most general reaction conditions relied on a slow regeneration of Cp_2TiCl in EtOAc as solvent at relatively low concentration of the starting material (0.02 M) in the presence of 20 mol % Cp_2TiCl_2, 2.0 equiv. of Zn, and 2.5 equiv. of Coll*HCl. However, in many cases the amounts of reagents used could be substantially reduced (10 mol % Cp_2TiCl_2, 0.2 equiv. of metal powder, and 0.5 equiv. of Coll*HCl).

The formation of THF derivatives through S_H2 reaction with mono- and disubstituted olefins was also investigated to define the overall scope of the transformation. Some of our results are summarized in Table 8. Not surprisingly, the monosubstituted alkene **53** gave essentially none of the desired **54** (5%). It is well known that the primary radicals produced during the 5-*exo* cyclization are rapidly trapped by Cp_2TiCl to yield the products of a reductive cyclization [17–20, 65, 66, 73, 74]. Epoxides containing disubstituted

Table 8 Mono- and disubstituted olefins as substrates in tetrahydofuran formation

Entry	Substrate	Product	Yield (%)
1	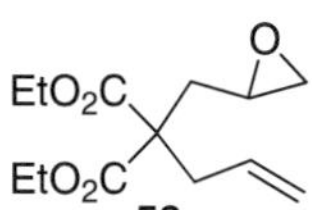 **53**	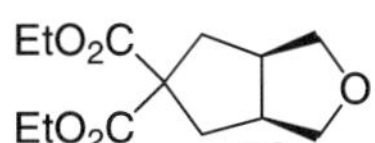 **54**	5
2	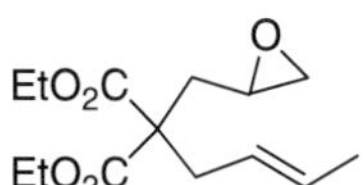 **55**	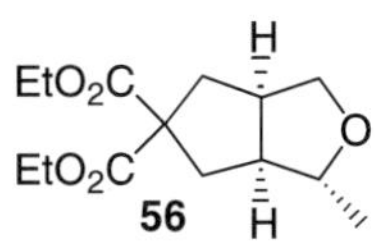 **56**	62, dr = 80:20
3	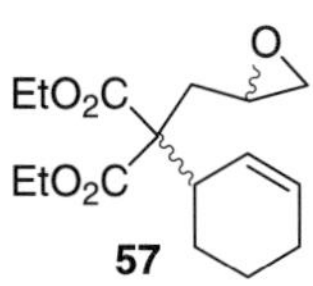 **57**	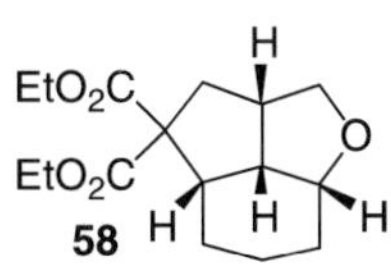 **58**	63
4	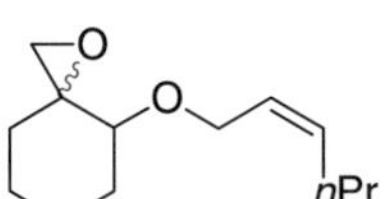 **59**	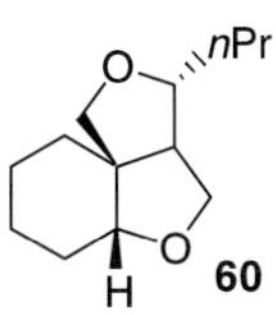 **60**	50[a], dr = 95:5

[a]dr (**59**) = 88 : 12

olefins constitute good substrates for the transformation. In the case of **55** (entry 2) a diastereoselectivity of 80 : 20 for **56** in favor of the isomer shown was observed. With a cyclic olefin, e.g., **57** (entry 3), which was employed as a roughly 50 : 50 mixture of diastereoisomers, the tricyclic product **58** was obtained as a single isomer.

Thus, the sole remaining stereocenter after epoxide opening controls the formation of three other stereocenters. It should be noted that the synthesis of enantiomerically pure substrates via palladium-catalyzed allylic alkylation [80] is possible and offers an access to the products in enantiomerically pure form. This possibility and the diastereoconvergent course of our reaction are extremely attractive for the synthesis of complex molecules.

The synthesis of **60** from **59** is of interest for two reasons. Synthetically, a structurally very complex molecule is accessible in a single step from a simple precursor. From a mechanistic point of view, the reaction proved very valuable because all by-products could be isolated and characterized. They are shown in Scheme 18.

The by-products originate from the initial 5-*exo* cyclization and constitute the product of the usual reductive cyclization and protonation of the Ti – O and Ti – C bonds **61** (dr = 71 : 29) and the product **62** (single isomer) of β-hydride elimination from an organotitanium intermediate in a ratio of 63 : 37 and in a combined yield of 35%. The major isomer of **61** was assigned as all-*cis*, in analogy to related systems that reacted with essentially the same selectivity. The mixture of these compounds can be converted into **61** quantitatively by hydrogenation over Pd/C. It turned out that **62** was converted into

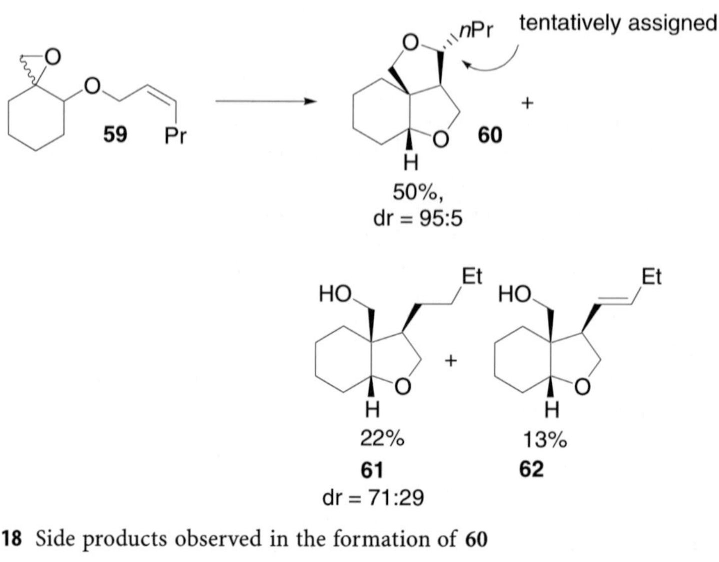

Scheme 18 Side products observed in the formation of **60**

the major isomer of **61**. Therefore, the structure of **62** was also assigned as all-*cis*.

These findings imply that, in the case of secondary radicals, the trapping with a second equivalent of Cp_2TiCl can compete with tetrahydrofuran formation, and that β-hydride elimination can kinetically compete with protonation of Ti – C bonds under our protic conditions [65, 66, 73, 74]. It remains to be seen how the reaction can be completely driven toward the tricyclic system **60** by ligand variation of the catalyst.

5
Conclusion

In summary, we have gained a comprehensive picture of titanocene-mediated epoxide opening through electron transfer. The investigations were carried out by collaboration between synthetic organic, physical organic, and computational chemistry, which has resulted in unique insights into the reaction at a molecular level. Moreover, by using these results we were able to develop novel reactions and catalysts, and will be able to do so in the future.

Acknowledgements We are grateful for continued financial support by the Danish Natural Science Research Council, the Deutsche Forschungsgemeinschaft, the Alexander von Humboldt-Stiftung, and the Fonds der Chemischen Industrie

References

1. Lane BS, Burgess K (2003) Chem Rev 103:2457
2. McGarrigle EM, Gilheany DC (2005) Chem Rev 105:1564
3. Jacobsen EN (2000) Acc Chem Res 33:421
4. Gansäuer A, Bluhm H (2000) Chem Rev 100:2771
5. Gansäuer A, Lauterbach T, Narayan S (2003) Angew Chem Int Ed 42:5556
6. Birch AJ (1949) J Proc R Soc NSW 83:245
7. Hallsworth AS, Henbest HB (1957) J Chem Soc 4604
8. Brown HC, Ikegami S, Kawakami JH (1970) J Org Chem 35:3243
9. Benkesser RA, Rappa A, Wolsieffer LA (1986) J Org Chem 51:3391
10. Bartmann E (1986) Angew Chem Int Ed 25:653
11. Dorigo AE, Houk KN, Cohen T (1989) J Am Chem Soc 111:8976
12. Conrow RE (1993) Tetrahedron Lett 34:5533
13. Bachki A, Foubeto F, Yus M (1995) Tetrahedron Asymmetry 6:1907
14. Bachki A, Foubeto F, Yus M (1996) Tetrahedron Asymmetry 7:2997
15. Cox JD, Pilcher G (eds) (1970) Thermochemistry of organic and organometallic compounds. Academic, London
16. Seeman JI (1983) Chem Rev 83:83
17. Nugent WA, RajanBabu TV (1998) J Am Chem Soc 110:8561
18. RajanBabu TV, Nugent WA (1989) J Am Chem Soc 111:4525
19. RajanBabu TV, Nugent WA, Beattie MS (1990) J Am Chem Soc 112:6408
20. RajanBabu TV, Nugent WA (1994) J Am Chem Soc 116:986

21. Halgren TH, Roberts JD, Horner JH, Martinez F, Tronche C, Newcomb M (2000) J Am Chem Soc 122:2988
22. Friedrich J, Dolg M, Gansäuer A, Geich-Gimbel D, Lauterbach T (2005) J Am Chem Soc 127:7071
23. Sharpless KB, Umbreit MA, Nieh MT, Flood TC (1972) J Am Chem Soc 94:6538
24. Su H, Walder L, Zhang Zd, Scheffold R (1988) Helv Chim Acta 71:1073
25. Bonhote P, Scheffold R (1991) Helv Chim Acta 74:1425
26. Gansäuer A, Rinker B (2002) Tetrahedron 58:7017
27. Matsukawa M, Tabuchi T, Inanaga J, Yamaguchi M (1987) Chem Lett 2101
28. Manzer LE (1982) Inorg Synth 21:135
29. Freudenberger JH, Konradi AW, Pedersen SF (1989) J Am Chem Soc 111:8014
30. Konradi AW, Kemp SJ, Pedersen SF (1994) J Am Chem Soc 116:1316
31. Konradi AW, Pedersen SF (1992) J Org Chem 57:28
32. Reetz MT, Griebenow N (1996) Liebigs Ann Chem 335
33. Kammermeier B, Beck G, Holla W, Jacobi D, Napierski B, Jendralla H (1996) Chem Eur J 2:307
34. Kochi JK, Singleton DM, Andrews LJ (1968) Tetrahedron 24:3503
35. Molander GA (2001) In: Renaud P, Sibi MP (eds) Radicals in organic synthesis, vol 1. Wiley, Weinheim, p 153 and references cited therein
36. Hasegawa E, Curran DP (1993) Tetrahedron Lett 34:1717
37. Shabangi M, Kuhlman ML, Flowers RA II (1999) Org Lett 1:2133
38. Enemærke RJ, Hertz T, Skrydstrup T, Daasbjerg K (2000) Chem Eur J 6:3747
39. Beckwith ALJ, Schiesser C (1985) Tetrahedron 41:3925
40. Schiesser CH, Wild LM (1996) Tetrahedron 42:13265
41. Walton JC (1998) Acc Chem Res 31:99
42. Curran DP, Fevig TL, Jasperse CP, Totleben MJ (1992) Synlett 943
43. Krief A, Laval AM (1999) Chem Rev 99:745
44. Wild FRW, Zsolnai L, Huttner G, Brintzinger HH (1982) J Organomet Chem 232:233
45. Collins S, Kuntz BA, Taylor NJ, Ward DG (1988) J Organomet Chem 342:21
46. Jaquith JB, Guan J, Wang S, Collins S (1995) Organometallics 14:1079
47. Cesarotti E, Kagan HB, Goddard R, Krüger C (1978) J Organomet Chem 162:297
48. Pedersen SU, Daasbjerg K (2001) In: Balzani V (ed) Electron transfer in chemistry, vol 1. Wiley, Weinheim, p 422
49. Enemærke RJ, Larsen J, Skrydstrup T, Daasbjerg K (2004) Organometallics 23:1866
50. Enemærke RJ, Larsen J, Skrydstrup T, Daasbjerg K (2004) J Am Chem Soc 126:7853
51. Enemærke RJ, Larsen J, Hjøllund GH, Skrydstrup T, Daasbjerg K (2005) Organometallics 24:1252
52. Mendes P, Gepasi, version 3.21. Virginia Tech, Blacksburg, VA. http://www.gepasi.org
53. Sekutowski DJ, Stucky GD (1975) Inorg Chem 14:2192
54. Sekutowski DJ, Jungst R, Stucky GD (1978) Inorg Chem 17:1848
55. Negishi EI (1999) Chem Eur J 5:411
56. Ahlrichs R, Bär M, Baron HP, Bauernschmitt R, Böcker S, Ehrig M, Eichkorn K, Elliott S, Furche F, Haase F, Häser M, Horn H, Huber C, Huniar U, Kattannek M, Kölmel C, Kollwitz M, May K, Ochsenfeld C, Öhm H, Schäfer A, Schneider U, Treutler O, von Arnim M, Weigend F, Weis F, Weiss H (2000) TURBOMOLE Version 5.3, Universität Karlsruhe
57. Becke AD (1988) Phys Rev A 38:3098
58. Schäfer A, Horn H, Ahlrichs R (1992) J Chem Phys 97:2571
59. Gansäuer A, Lauterbach T, Bluhm H, Noltemeyer M (1999) Angew Chem Int Ed 38:2909

60. Gansäuer A, Bluhm H, Rinker B, Narayan S, Schick M, Lauterbach T, Pierobon M (2003) Chem Eur J 9:531
61. Gansäuer A, Rinker B, Pierobon M, Grimme S, Gerenkamp M, Mück-Lichtenfeld C (2003) Angew Chem Int Ed 42:3687
62. Gansäuer A, Rinker B, Ndene-Schiffer N, Pierobon M, Grimme S, Gerenkamp M, Mück-Lichtenfeld C (2004) Eur J Org Chem :2337
63. Mudryk B, Cohen T (1991) J Am Chem Soc 113:1866
64. Malacria M (1996) Chem Rev 96:289
65. Gansäuer A, Pierobon M, Bluhm H (1998) Angew Chem Int Ed 37:101
66. Gansäuer A, Bluhm H, Pierobon M (1998) J Am Chem Soc 120:12849
67. Berno P, Floriani C, Chiesi-Villa A, Guastini C (1990) Organometallics 9:1995
68. Hortmann K, Diebold J, Brintzinger HH (1993) J Organomet Chem 445:107
69. Curtis MD, Thanedar S, Butler WM (1984) Organometallics 3:1855
70. Spellmeyer DC, Houk KN (1987) J Org Chem 52:959
71. Isaacs N (1995) Physical organic chemistry, 2nd edn. Longman, Harlow, p 418
72. Hammond GS (1955) J Am Chem Soc 77:334
73. Gansäuer A, Pierobon M (2000) Synlett 1357
74. Gansäuer A, Pierobon M, Bluhm H (2001) Synthesis 2500
75. Barrero AF, Rosales A, Cuerva JM, Oltra E (2003) Org Lett 5:1935
76. Justicia J, Rosales A, Buñuel E, Oller-López JL, Valdivia M, Haïdour A, Oltra JE, Barrero AF, Cárdenas DJ, Cuerva JM (2004) Chem Eur J 10:1778
77. Schreiber SL (2000) Science 287:1964
78. Lee D, Sello JK, Schreiber SL (2000) Org Lett 2:709
79. Bailey WF, Longstaff SC (2001) Org Lett 3:2217
80. Trost BM, Crawley ML (2003) Chem Rev 103:2921

Top Curr Chem (2006) 263: 71–106
DOI 10.1007/128_030

Published online: 16 February 2006

Tin-Free Radical Reactions Mediated by Organoboron Compounds

Vincent Darmency · Philippe Renaud (✉)

Departement für Chemie und Biochemie, Universität Bern, Freiestrasse 3, 3012 Bern, Switzerland
philippe.renaud@ioc.unibe.ch

Abstract The development of tin-free methods to run radical reactions is crucial for their applications in industrial processes as well as in drug discovery projects. Within the last 10 years, organoboron compounds have demonstrated that they represent one of the most attractive approach to substitute tin reagents in radical process. This review summarizes the results obtained with organoboron compounds as a source of alkyl radicals. Four different strategies are described, i.e., the use of organoboranes as radical initiators, as a source of alkyl radicals, as chain transfer reagents, and finally as reducing agents.

Keywords Atom transfer reactions · Boron · C – C bond formation · Conjugate addition · Radical initiators · Radical reaction · Tin-free

1 Introduction

The ability of organoboron compounds to participate in free radical reactions has been identified since the earliest investigation of their chemistry [1–3]. For instance, the autoxidation of organoboranes (Scheme 1) has been proven to involve radical intermediates [4, 5]. This reaction has led recently to the use of triethylborane as a universal radical initiator functioning under a very wide range of reaction conditions (temperature and solvent) [6, 7].

Interestingly, homolytic substitution at boron does not proceed with carbon centered radicals [8]. However, many different types of heteroatom centered radicals, for example alkoxyl radicals, react efficiently with the organoboranes (Scheme 2). This difference in reactivity is caused by the Lewis base character of the heteroatom centered radicals. Indeed, the first step of the homolytic substitution is the formation of a Lewis acid-Lewis base complex between the borane and the radical. This complex can then undergo a β-fragmentation leading to the alkyl radical. This process is of particular interest for the development of radical chain reactions.

Our review of the use of organoboron compounds in radical chemistry will concentrate on applications where the organoborane is used as an initiator, as a direct source of carbon-centered radicals, as a chain transfer reagent and finally as a radical reducing agent. The simple formation of carbon-heteroatom bonds via a radical process is not treated in this review since it has been treated in previous review articles [3, 9].

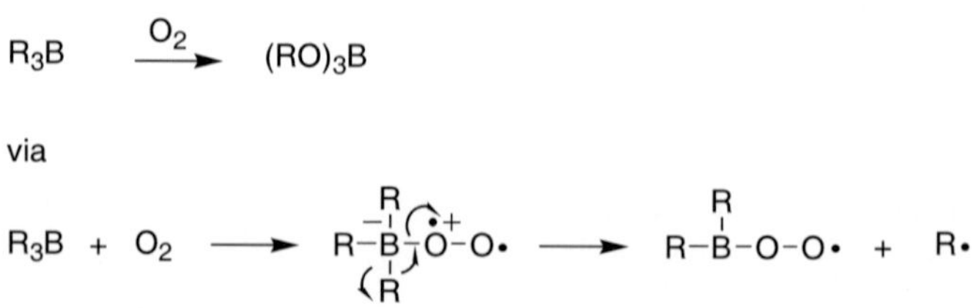

Scheme 1 Autoxidation of organoboranes

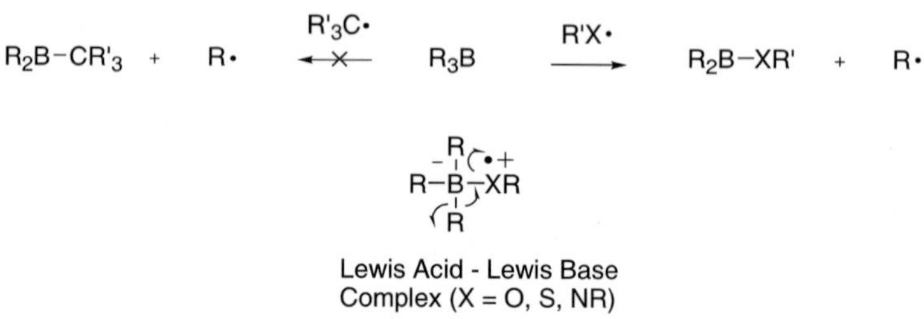

Scheme 2 Reactivity of carbon- and heteroatom-centered radicals towards organoboranes

2
Organoboranes as Radical Initiators

Utimoto and Oshima were the first to apply the reaction of triethylborane with oxygen to initiate radical reactions [6, 10]. Over classical initiators, the system Et_3B/O_2 offers the great advantage of being efficient even at low temperature (– 78 °C). This aspect proved to be particularly important for the development of stereoselective radical reactions and for radical reactions involving thermally unstable adducts or products. Review articles describing the use of triethylborane as a radical initiator have appeared [3, 7]. The majority of the reported examples involves tin reagents and will not be discussed here. A few selected examples involving tin-free chemistry will be presented.

2.1
In Reductive Processes

2.1.1
Reduction of Halides and Related Compounds

Triethylborane can initiate the formation of silyl or germanyl radical from the related hydrides. For example, Evans reported the cyclization of acyl radicals to vinylogous carbonates in the presence of $(TMS)_3SiH$. By using Et_3B/O_2 initiation rather than AIBN, the *cis*-oxepanones are obtained in higher stereoselectivity and yield since the decarbonylation of the intermediate acyl radical is suppressed (Scheme 3, Eq. 3a) [11]. In another striking example, a thermally unstable propargyl bromide cobalt complex cyclizes in the presence of Ph_2SiH_2 under Et_3B/O_2 initiation at 20 °C. A mixture of reduced and bromine atom transfer products are isolated (Eq. 3b) [12].

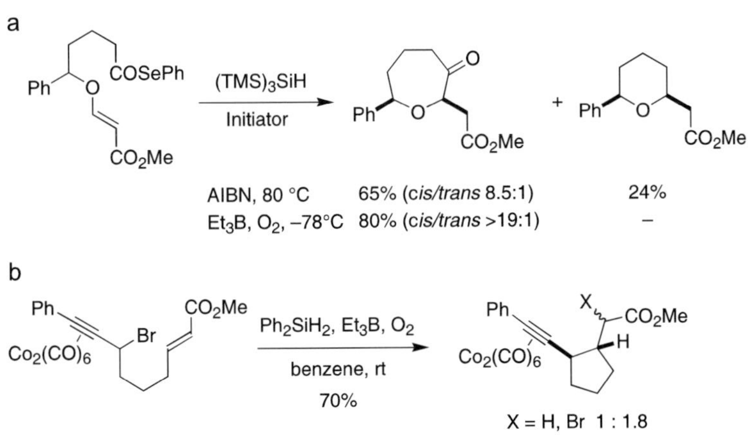

Scheme 3 Reductive cyclizations with silanes

Interestingly, Et_3B/O_2 initiation can be performed in aqueous solution [13]. For instance, a wide range of aryl and alkyl halides are reduced in water by water-soluble organosilanes using Et_3B/O_2 initiation (Scheme 4) [14].

Recently, Curran described a procedure using triethylborane for the synthesis of spirooxindoles and spirodihydroquinolones through intramolecular addition of aryl radicals at the *ipso* position 4-alkoxy-substituted aromatic rings [15]. The key step for a formal synthesis of the vasopressin inhibitor SR121463A is described in Scheme 5. The initiation was performed with Et_3B in an open to air reaction vessel.

Germanes are also used for the reduction of various organic halides at ambient temperature under Et_3B/O_2 initiation. For example, tri-2-furylgermane mediated radical cyclizations of aryl iodides proceed in good yields (Scheme 6, Eq. 6a) and are also possible with $NaBH_4$ in the presence of a catalytic amount of triphenylgermane (Eq. 6b) [16].

Tin-free radical reduction by an organophosphite [17] and phosphinic acid can also be initiated by Et_3B/O_2. Radical cyclizations using phosphinic acid neutralized with sodium carbonate and Et_3B/O_2 as a radical initiator

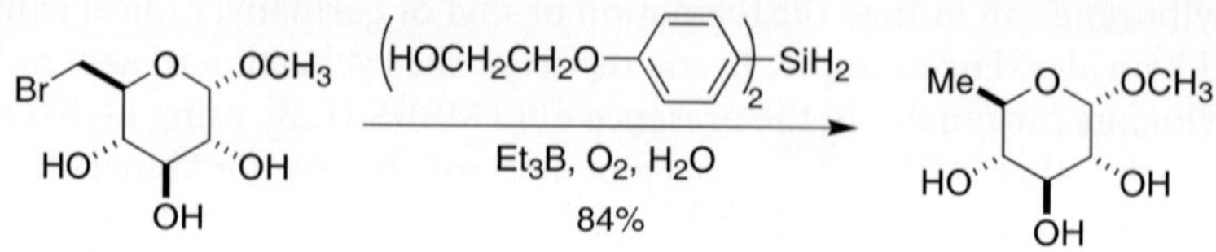

Scheme 4 Silane mediated reduction of a bromide in water

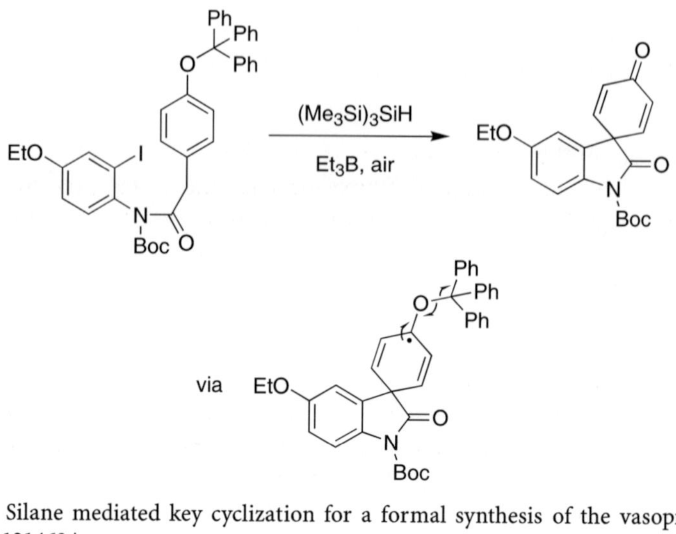

Scheme 5 Silane mediated key cyclization for a formal synthesis of the vasopressin inhibitor SR121463A

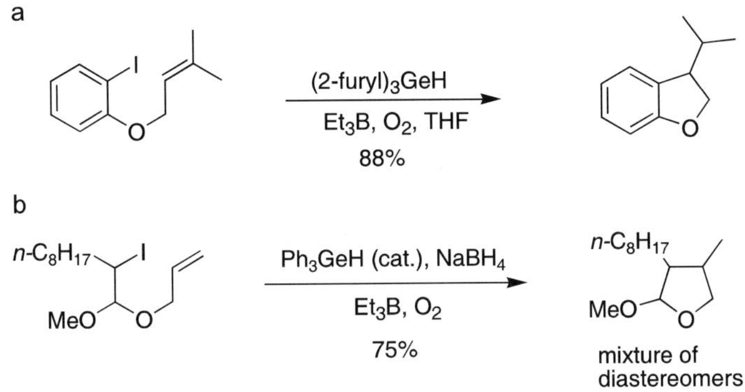

Scheme 6 Triarylgermane mediated radical cyclizations

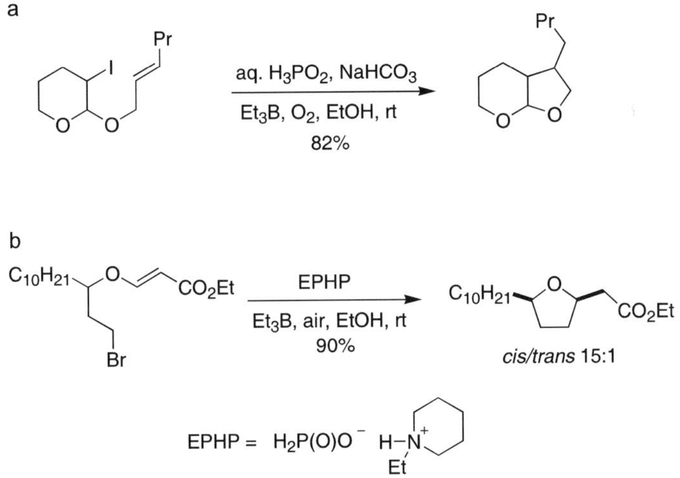

Scheme 7 Phosphinic acid mediated radical cyclizations at room temperature

in aqueous ethanol were recently reported (Scheme 7, Eq. 7a) [18]. A similar stereoselective cyclization of a β-alkoxyacrylate with 1-ethylpiperidinium hypophosphite (EPHP) at room temperature was described by Lee (Eq. 7b) [19].

2.1.2
Reductive Addition of Heteroatom Centered Radicals to Alkynes and Alkenes

Tris(trimethylsilyl)silane [20, 21], thiols [22], germanes [23–25] and gallium hydride [26] can be added easily to terminal alkynes in the presence of Et_3B/O_2. This process was extended to internal alkenes (Scheme 8, Eq. 8a) as well as silyl enol ethers (Eq. 8b) by using tri-2-furylgermane. In this last case, basic or acidic treatment of the main *syn* β-siloxygermane furnishes the corresponding *E*- or *Z*-alkene, respectively [24].

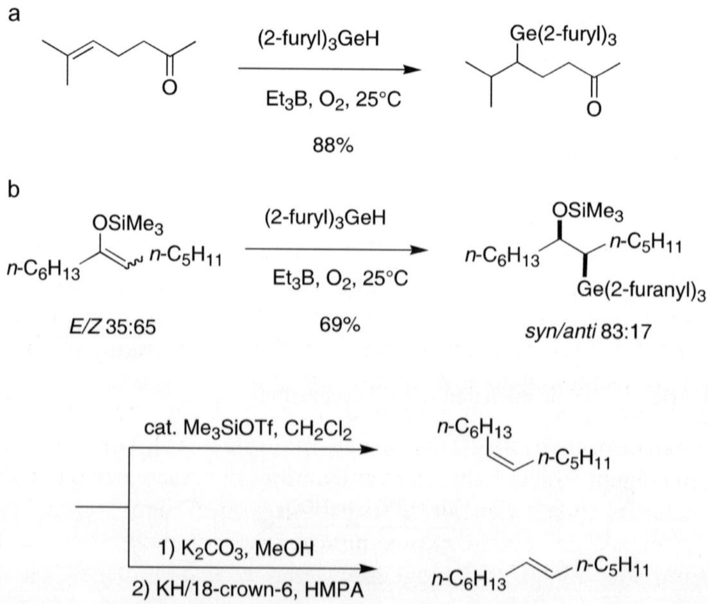

Scheme 8 Addition of triarylgermanes to alkenes and enol silanes

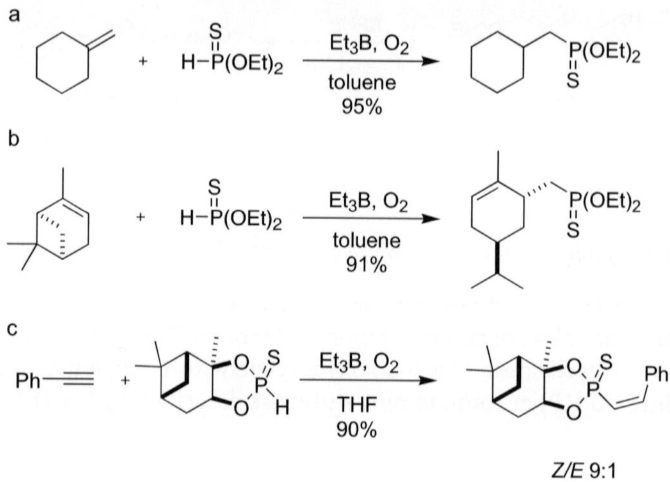

Scheme 9 Reductive processes with phosphorus based reagents

The addition of hypophosphites to alkenes under Et_3B initiation is also reported [27]. Piettre described recently the addition of diethylthiophosphite to alkenes leading to the formation of thiophosphonates (Scheme 9, Eq. 9a) [28]. Interestingly, this reaction can be used for cyclization of dienes and ring opening of strained alkenes such as α-pinene (Eq. 9b). Parson prepared an alkenyl thiophosphonate by reaction of a chiral thiophosphite with phenylacetylene (Eq. 9c) [29].

2.2 In Fragmentation Processes

Oshima [30] reported a radical alkenylation of α-halo carbonyl compounds under mild conditions by utilizing alkenylindium reagents. Using 0.5 equivalent of triethylborane as a radical initiator at ambient temperature, we demonstrated that this process affords the alkenylation products in high yield (Scheme 10, Eq. 10a). Styrylation reaction showed retention of the stereochemistry from starting alkenylindium (Eq. 10b).

An allylzirconium reagent, prepared from Cp_2ZrCl_2 and allylmagnesium chloride, can be used for the allylation of α-iodoesters [31]. The reaction of allyl zirconium reagent with α-halo esters and amides in presence of triethylborane provides a useful and efficient alternative for organotin chemistry (Scheme 11). This reaction has been extended to a three component coupling process. Similar reactions with allylgallium reagent in water are also reported [26].

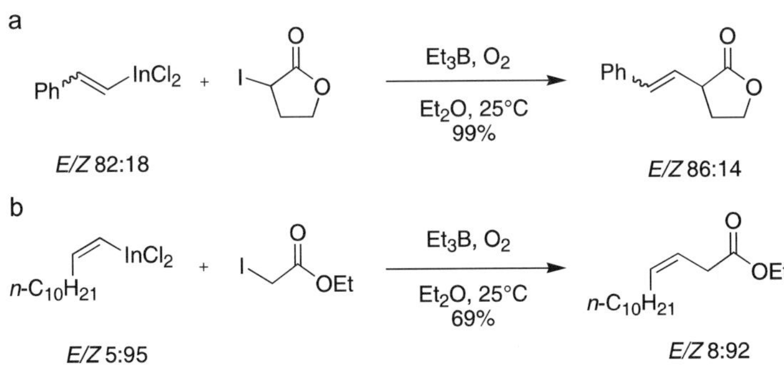

Scheme 10 Alkenylation reactions with alkenylindium derivatives

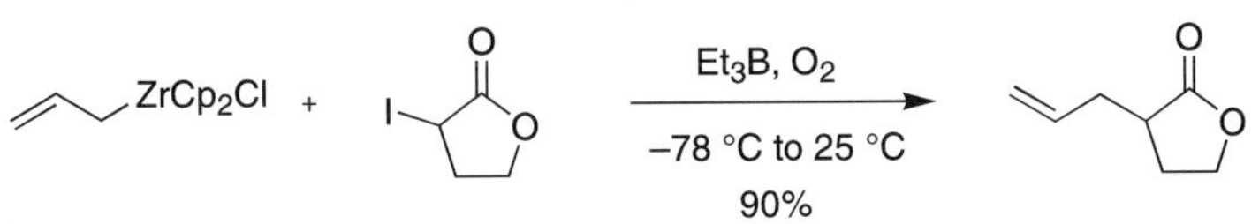

Scheme 11 Allylation with allylzirconium reagents

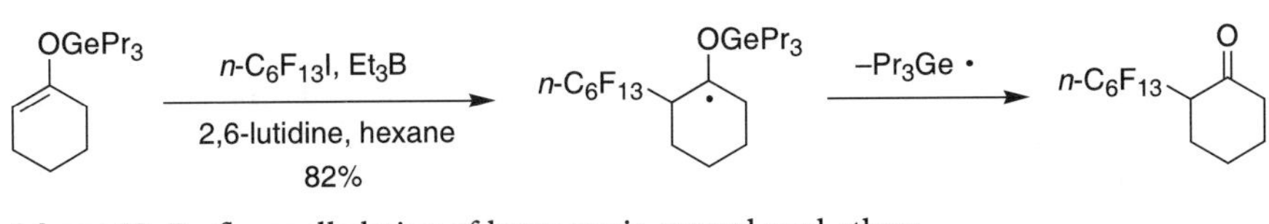

Scheme 12 Perfluoroalkylation of ketones via germyl enol ethers

Germyl enol ethers react with perfluoroalkyl iodides under Et_3B initiation to give α-perfluoroalkyl ketones. The intermediate radical adduct decomposes readily via β-elimination and provides the α-perfluoroalkyl ketone and a trialkylgermanyl radical as a chain carrier (Scheme 12) [32].

2.3 In Atom Transfer Processes

2.3.1 Iodine Atom Transfer

Triethylborane in combination with oxygen provides an efficient and useful system for iodine atom abstraction from alkyl iodide, and thus is a good initiator for iodine atom transfer reactions [13, 33, 34]. Indeed, the ethyl radical, issued from the reaction of triethylborane with molecular oxygen, can abstract an iodine atom from the radical precursor to produce a radical R˙ that enters into the chain process (Scheme 13). The iodine exchange is fast and efficient when R˙ is more stable than the ethyl radical.

Et_3B-induced addition of perfluoro alkyliodides [35], α-iodoesters (Scheme 14, Eq. 14a) [36], iodoamides [37], α-iodonitriles [36], and simple

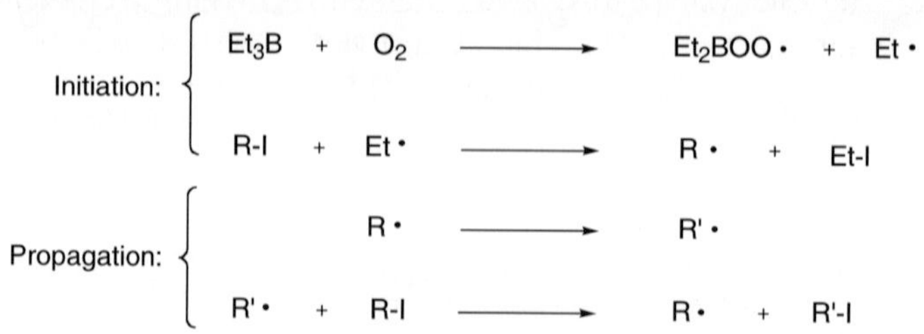

Scheme 13 Mechanism of the Et_3B-mediated iodine atom transfer reaction

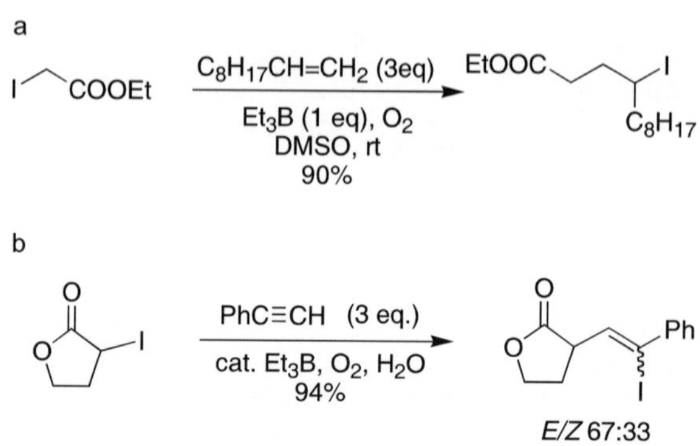

Scheme 14 Intermolecular additions through iodine atom transfer

alkyl iodides [38] to alkenes and alkynes have been reported. Interestingly, these reactions were also performed with success in aqueous media [13, 39, 40] demonstrating the ability of Et_3B to act as initiator in water (Eq. 14b) [41].

Triethylborane is also an excellent initiator for intramolecular iodine atom-transfer reactions. For example, cyclization of the propargyl α-iodo-acetal depicted in Scheme 15 (Eq. 15a) gives the corresponding bicylic vinyliodide in high yield [38]. Allyl iodoacetamides (Eq. 15b) and allyl iodoacetates (Eq. 15c) cyclize cleanly under Et_3B/O_2 initiation. In the case of the ester, the reaction has to be run in refluxing benzene in order to allow *Z*/*E*-ester isomerization prior to cyclization [42, 43]. No trace of cyclized product is detected when the reaction is carried out at room temperature. Interestingly, by running the same reaction in water, Oshima obtained the desired lactone in 78% yield. It was suggested that water facilitates the *Z*/*E* isomerization. Efficient preparation of medium and large ring lactones in water have also been reported [39, 40].

Examples of tandem intermolecular addition-cyclization under iodine atom-transfer conditions are depicted in Scheme 16 [38, 41].

Et_3B-induced radical cascade reactions with 1,5-enynes and 1,5-diynes have been applied to the synthesis of dioxatriquinanes and tricyclic glucoconjugates (Scheme 17) [44, 45]. Some of these elegant cascade cyclizations were also performed under mild conditions at – 50 °C.

Using acyclic and cyclic *N*-tosylated iodomethylaziridines, Taguchi investigated annulation reactions [46]. The reaction with electron-rich alkenes such as enol ethers proceeds smoothly as illustrated in Scheme 18.

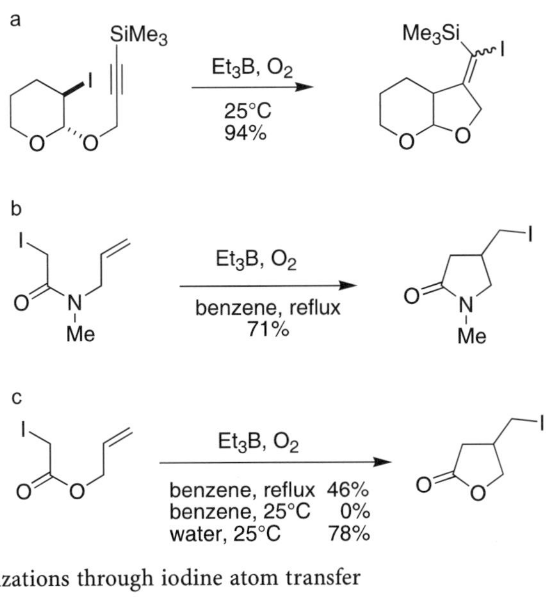

Scheme 15 Cyclizations through iodine atom transfer

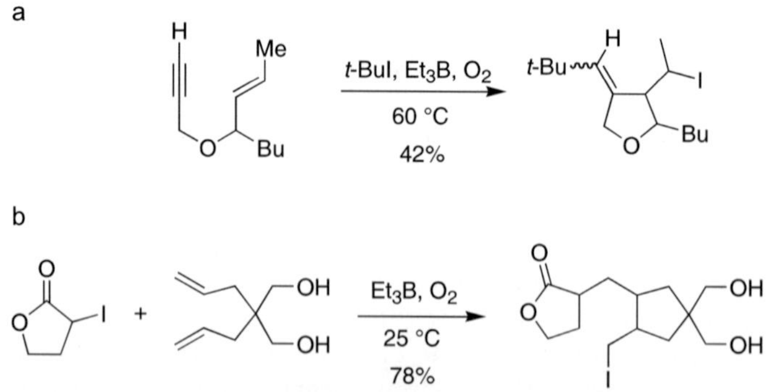

Scheme 16 Tandem intermolecular addition-cyclization reactions

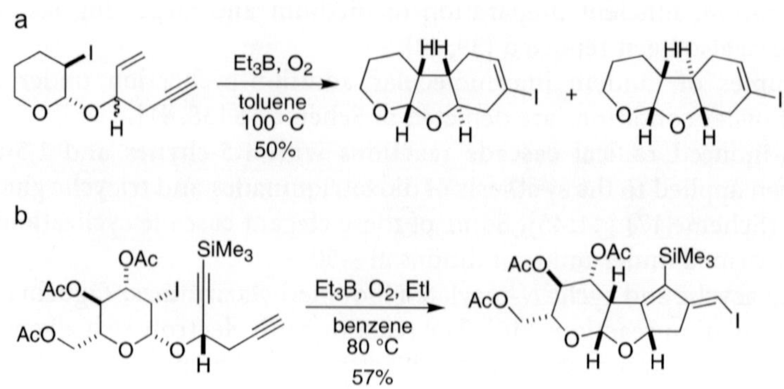

Scheme 17 Cascade cyclizations

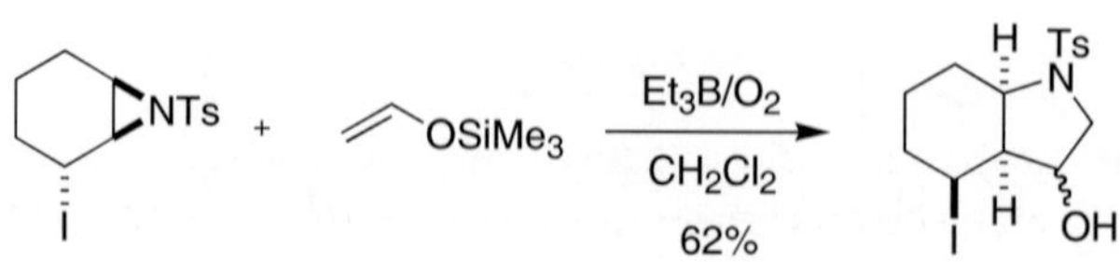

Scheme 18 Preparation of pyrrolidine derivatives via annulation with iodomethylaziridine derivatives

Silyl enol ethers have also been used as a trap for electrophilic radicals derived from α-haloesters [36] or perfluoroalkyl iodides [32]. They afford the α-alkylated ketones after acidic treatment of the intermediate silyl enol ethers (Scheme 19, Eq. 19a). Similarly, silyl ketene acetals are converted into α-perfluoroalkyl esters upon treatment with perfluoroalkyl iodides [32, 47]. The Et_3B/O_2-mediated diastereoselective trifluoromethylation [48, 49] (Eq. 19b) and (ethoxycarbonyl)difluoromethylation [50, 51] of lithium enolates derived from *N*-acyloxazolidinones have also been achieved. More recently, Mikami [52] succeeded in the trifluoromethylation of ketone enolates

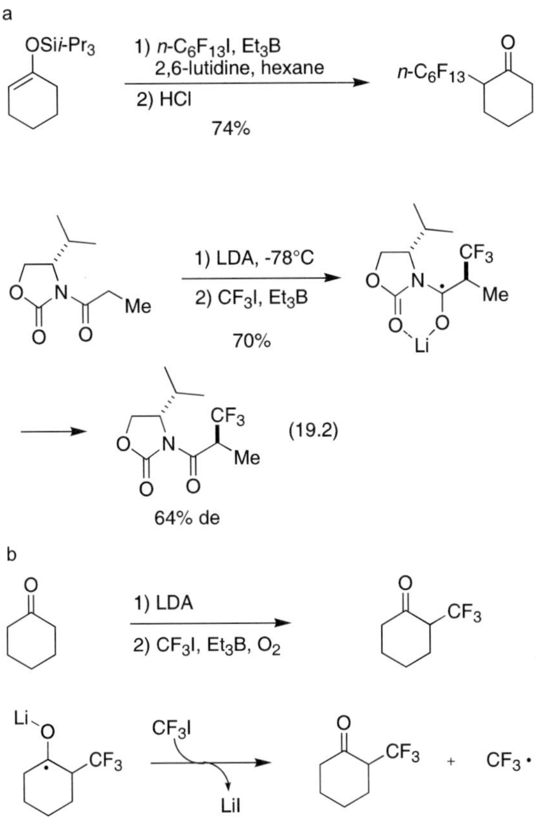

Scheme 19 Perfluoroalkylation of silyl enol ethers and lithium enolates via iodine atom transfer

(Eq. 19c). The mechanism of this transformation involves either a final iodine transfer step or an electron transfer process that give back the trifluoromethyl radical.

2.3.2 Bromine Atom Transfer

Bromides are less reactive than the corresponding iodides in atom transfer processes. However, activated bromides such as diethyl bromomalonate [36] and bromomalonitrile [53] react with olefins under Et_3B/O_2 initiation. Kharasch type reactions of bromotrichloromethane with alkenes are also initiated by Et_3B/O_2 [41]. On the other hand, a remarkable Lewis acid effect was reported by Porter. Atom-transfer reactions of an α-bromooxazolidinone amide with alkenes are strongly favored in the presence of Lewis acids such as $Sc(OTf)_3$ or $Yb(OTf)_3$, this reaction was successively applied to the

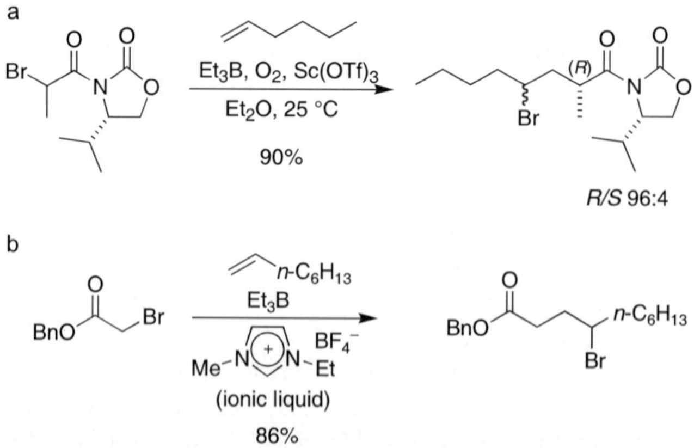

Scheme 20 Bromine atom transfer

diastereoselective alkylation of chiral oxazolidinone derivatives (Scheme 20, Eq. 20a) [54]. More recently, Oshima reported that bromine atom transfers take place at room temperature in ionic liquid media (Eq. 20b) [55].

2.3.3 Chlorine Atom Transfer

Radical [3 + 2] annulation involving *N*-allyl-*N*-chlorotosylamide provides a route to pyrrolidine derivatives (Scheme 21) [56].

Reaction of *N*-chlorosulfonyl derivatives with enol ether initiated by Et_3B has been reported (Scheme 22) [57]. The reaction mechanism involves a chlorine atom transfer followed by a Et_3N promoted elimination of HCl to produce stable enol ethers.

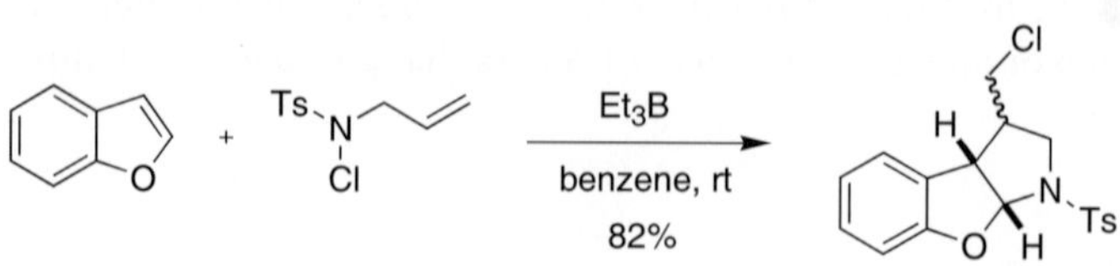

Scheme 21 Radical [3 + 2] annulation involving *N*-allyl-*N*-chlorotosylamide

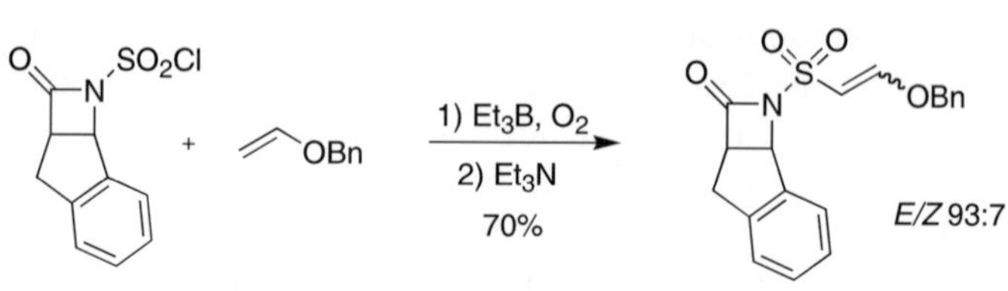

Scheme 22 Vinylation of β-lactamido *N*-sulfonyl chloride

3
Organoboron Compounds as a Source of Carbon-Centered Radicals

3.1
Conjugated Additions to Enones and Enals

One of the first synthetic applications of organoboranes in radical chemistry is the conjugate addition to enones (Scheme 23, Eq. 23a) and enals reported by Brown [58–61]. Addition to β-substituted enones and enals are not spontaneous and initiation with the oxygen [62], diacetyl peroxide [63], or under irradiation [63] is necessary (Eq. 23b). A serious drawback of this strategy is that only one of the three alkyl groups is efficiently transferred, so the method is restricted to trialkylboranes derived from the hydroboration of easily available and cheap alkenes. To overcome this limitation *B*-alkylboracyclanes have been used but this approach was not successful for the generation of tertiary alkyl radicals [64, 65].

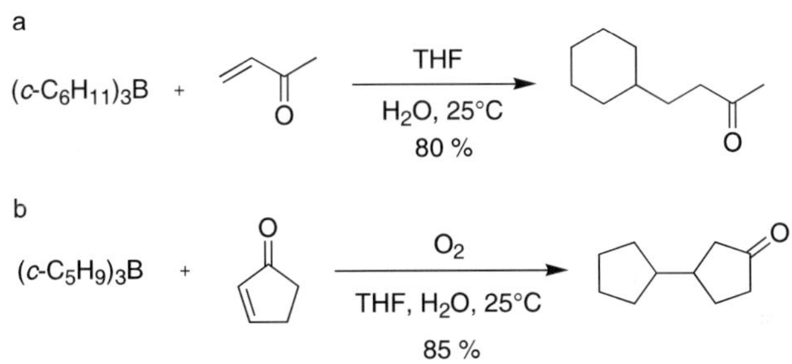

Scheme 23 The Brown conjugate addition

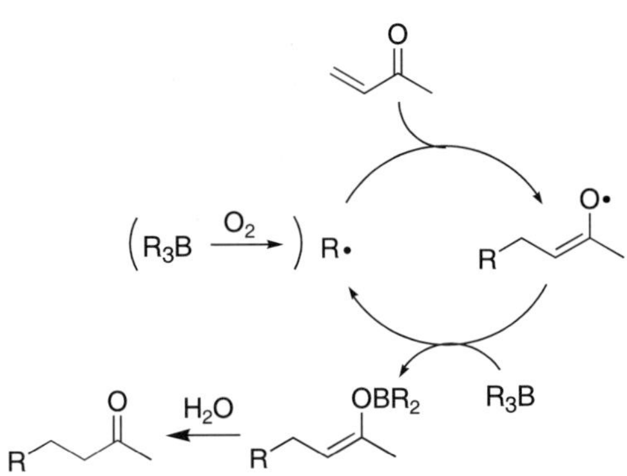

Scheme 24 Brown mechanism for the conjugate addition of organoboranes to methyl vinyl ketone

Toru has investigated the stereoselectivity of the conjugate addition of trialkylboranes to 2-arylsulfinylcyclopentenones. Excellent stereocontrol is achieved with different alkyl radicals (Scheme 27) [73–76]. In the acyclic series, the lack of diastereoselectivity in the addition step and a competitive Pummerer rearrangement have limited the synthetic potential of this reaction [77].

A serious drawback of the trialkylborane approach is the requirement to use a 1 : 1 ratio of trialkylborane/radical trap ratio to obtain good yields. Therefore, the method is restricted to trialkylboranes obtained by hydroboration of easily available and cheap alkenes. To overcome this limitation, *B*-alkylboracyclanes have been used [64, 65]. According to Brown and Negishi, 3,3-dimethylborinane, prepared from BH_3 and 2,4-dimethyl-1,4-pentadiene, is the most efficient reagent. With this system, a selective cleavage of the boron-alkyl bond is possible for secondary and tertiary alkyl groups (Scheme 28). This method, referred to later as the Brown-Negishi reaction, is not suitable for primary alkyl radicals (yield < 35%) and for radical traps substituted at the β-position. With these traps, the addition of extra oxygen is necessary to run the chain reaction and under these conditions the cleavage of the carbon-boron bond is not longer selective.

Recently, we have shown that similar results are obtained with cyclohexyldiethylborane (easily prepared from Et_2BH and cyclohexene). The effi-

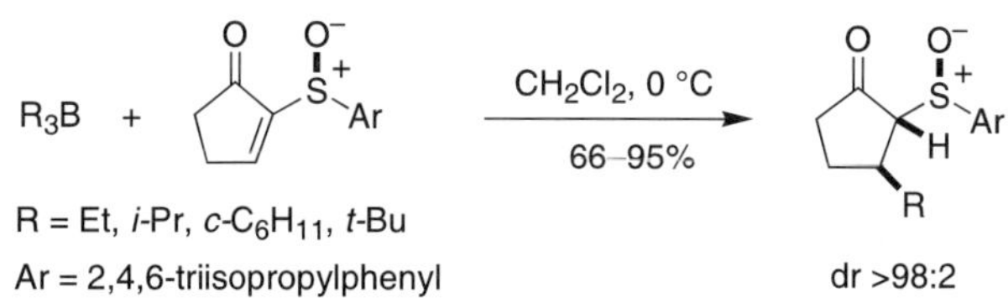

Scheme 27 Stereoselective addition to 2-arylsulfinylcyclopentenones

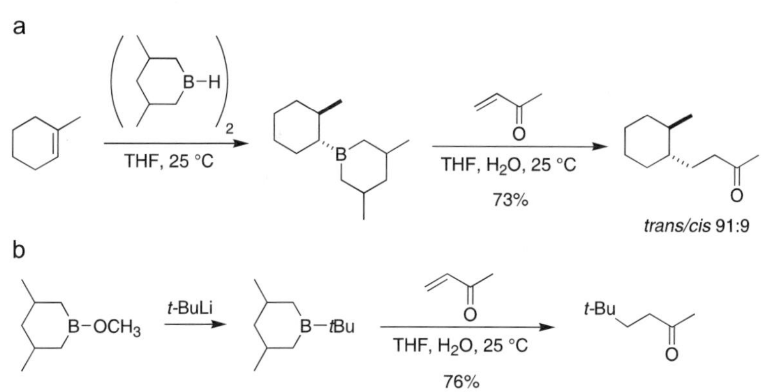

Scheme 28 Brown-Negishi reaction: selective formation of secondary and tertiary alkyl radicals

cient addition to methyl vinyl ketone is possible (Scheme 29, Eq. 29a). However, when cyclohexenone is used as radical trap, addition of oxygen is necessary and a 3 : 1 mixture of products resulting from the addition of cyclohexyl and ethyl radicals is obtained (Eq. 29b) [78].

In order to circumvent the lack of selectivity in the cleavage of trialkylboranes, *B*-alkylcatecholboranes can be used as precursor of alkyl radicals. They are extremely sensitive towards oxygen and they react readily with alkoxyl radicals. It was clearly demonstrated by ESR that the perboryl radical intermediate resulting from the complexation of *B*-methylcatecholborane with the alkoxyl radical is stabilized by delocalization onto the aromatic ring (Scheme 30) [79].

The observation of Davies and Roberts regarding the stability of the perboryl radical is at the origin of our own investigations about the use of *B*-alkylcatecholboranes as radical precursors. *B*-alkylcatecholboranes are expected to be more reactive than trialkylboranes and they are easily prepared from olefins via hydroboration with catecholborane with or without a catalyst. However, the most attractive feature of *B*-alkylcatecholboranes is the possibility to generate selectively one alkyl radical from an olefin; a possibility that trialkylboranes do not offer since no selective cleavage of the desired alkyl group is observed (vide supra). Indeed, reaction of *B*-alkylcatecholborane with a heteroatom centered radical leads in an irreversible manner to the alkyl radical since cleavage of the "wrong" B – O bond

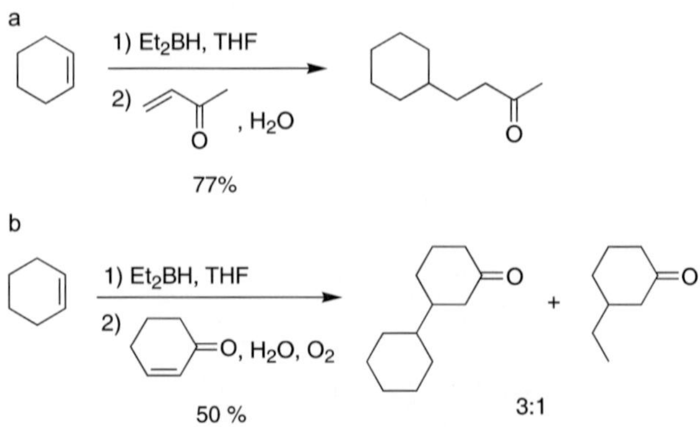

Scheme 29 Diethylboranes mediated conjugate addition of secondary alkyl radicals

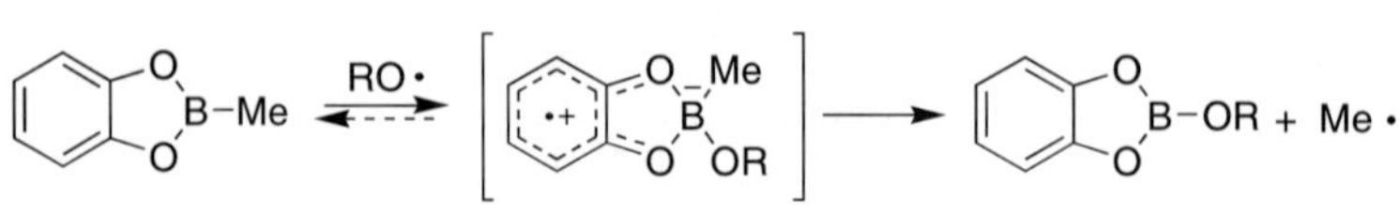

OR = OEt or O*t*-Bu

Scheme 30 Reaction of *B*-methylcatecholborane with alkoxyl radicals

is a reversible process that finally leads to the irreversible formation of an alkyl radical (Scheme 31) [9].

A modified version of the Brown-Negishi reaction using *B*-alkylcatecholboranes was reported (Scheme 32). This novel method is based on a simple one-pot procedure involving the hydroboration of various substituted alkenes with catecholborane, followed by treatment with catalytic amount of oxygen/DMPU/water and a radical trap. Efficient radical additions to α,β-unsaturated ketones and aldehydes have been reported. Primary alkyl radicals are efficiently generated by this procedure and the reaction has been applied to a 300 mmol scale synthesis of the γ-side chain of (–)-perturasinic

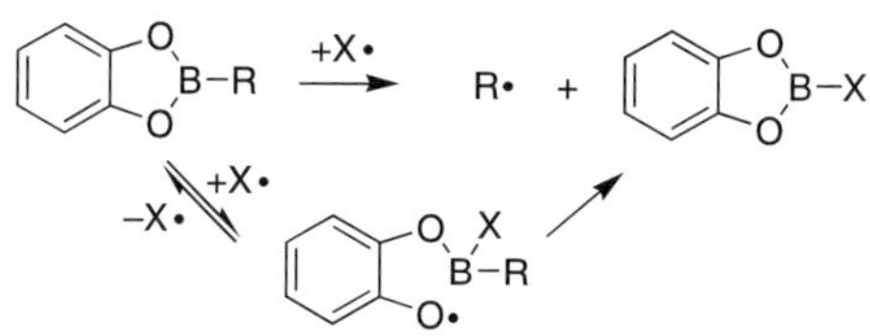

Scheme 31 Irreversible formation of alkyl radicals from *B*-alkylcatecholboranes

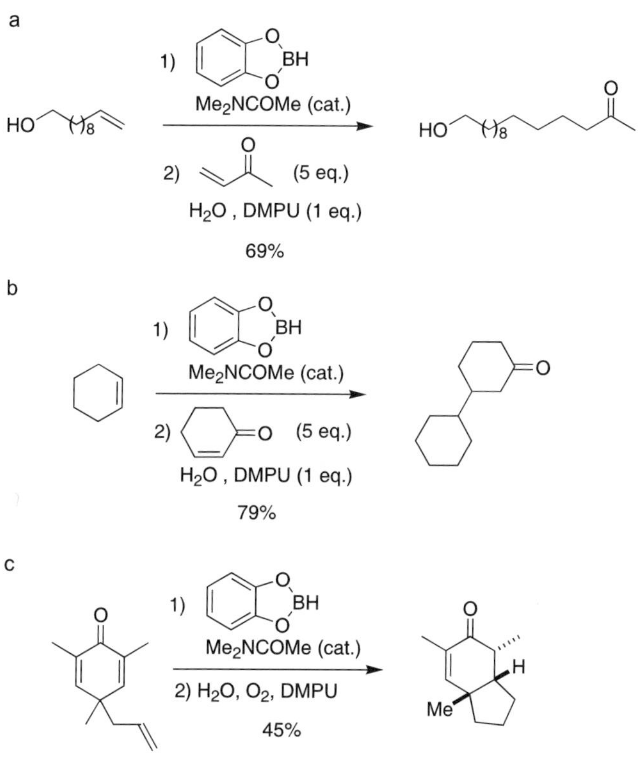

Scheme 32 *B*-Alkylcatecholborane mediated addition of organoboranes to enones

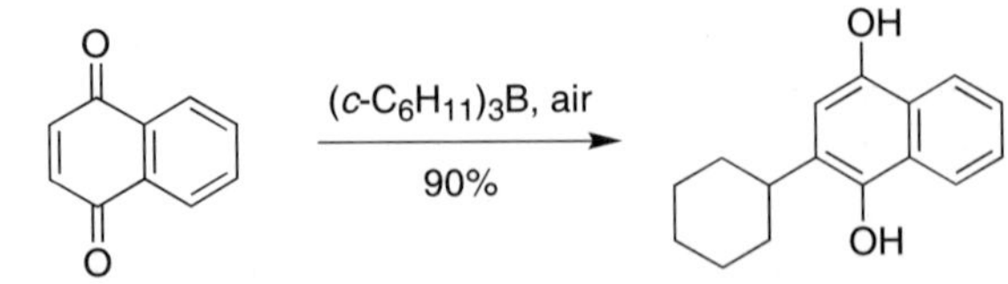

Scheme 33 Addition of alkyl radicals to quinones using trialkylboranes

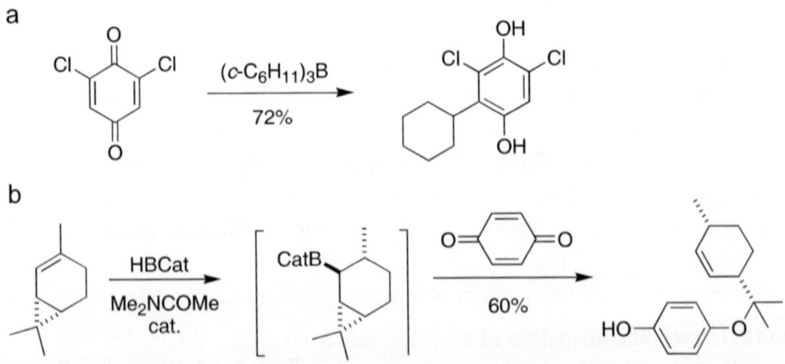

Scheme 34 Addition of *B*-alkylcatecholboranes to quinones: C- versus O-addition

acid (Eq. 32a) [80]. The reaction was also applied to the radical addition to cyclohexenone (Eq. 32b) and to other β-substituted enones and enals as well as to cyclization (Eq. 32c) and annulation reactions [78].

The reaction of trialkylboranes with 1,4-benzoquinones to give in quantitative yield 2-alkylhydroquinones was the first reaction of this type occurring without the assistance of a metal mediator [81,82]. An ionic mechanism was originally proposed but rapidly refuted since the reaction is inhibited by radical scavengers such as galvinoxyl and iodine [83]. This procedure is in many cases superior to the more widely use organometallic additions. For instance, when primary and secondary alkyl radicals have been used and afford the addition products in high yield (Scheme 33) [84].

The addition of *B*-alkylcatecholboranes to quinones has recently been investigated [85]. Good yield of the expected conjugate addition product are obtained with primary and most secondary radicals (Scheme 34, Eq. 34a). However, hindered secondary radicals and tertiary alkyl radicals afford an unexpected product resulting from a radical addition to the oxygen atom of the quinone (Eq. 34b).

3.2 Conjugate Addition to Activated Olefins

The modified Brown-Negishi and the *B*-alkylcatecholborane conjugate additions described above are limited to enone and enal radical traps. Other radical traps such as unsaturated esters, amides and sulfones fail to react under these conditions. This failure was interpreted as a consequence of an inefficient reaction of the radical adduct and *B*-alkylcatecholboranes. This inefficiency is caused by the insufficient density of unpaired electrons on the oxygen atom of theses radicals relative to ketone-enolate and aldehyde-enolate radicals. The use of a chain-transfer reagent which is able to convert a carbon-centered radical into an oxygen-centered radical allows one to solve this problem. The Barton carbonate PTOC-OMe (PTOC = pyridine-2-thione-*N*-oxycarbonyl) [86, 87] proved to be an excellent radical chain transfer reagent according to Scheme 35 (Eq. 35a) [88]. Interestingly, the same reagent proved to be an excellent initiator under irradiation with a standard 150 W tungsten lamp (Eq. 35b). The PTOC-OMe is a stable reagent easily obtained by the reaction of the commercially available sodium salt of *N*-hydroxypyridine-2-thione with methyl chloroformate. A related strategy has been developed by Dalko and Cossy using the Barton ester PTOC-Ph as chain transfer reagent [89].

In a preliminary study, in situ generated *B*-alkylcatecholboranes were allowed to react with PTOC-OMe under irradiation with a standard 150 W lamp. The *S*-pyridyl products coming from primary, secondary and tertiary alkyl radicals were isolated in moderate to good yields [88]. Based on these initial results, a procedure for conjugate addition to various activated alkenes was developed. A one-pot procedure involving hydroboration of an alkene with catecholborane followed by irradiation in the presence of five equivalents of an activated alkene and three equivalents of the chain transfer reagent PTOC-OMe was developed (Scheme 36) [88].

In contrast to the tin hydride-mediated reaction (Giese reaction) [90], no slow addition of the chain carrier is necessary. This is easily understandable

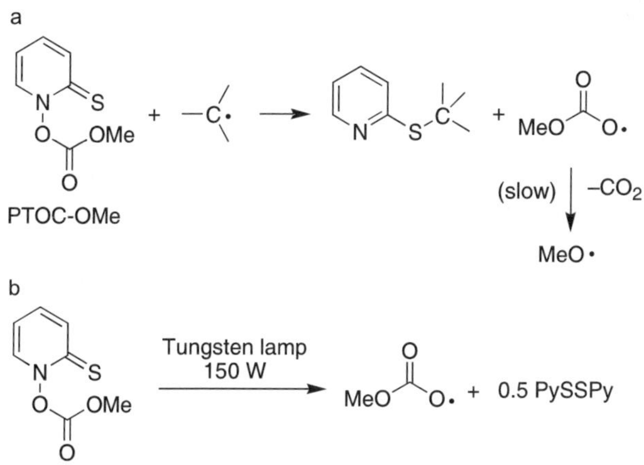

Scheme 35 Barton carbonate PTOC-OMe, a radical chain transfer reagent able to convert a C-centered radical into an O-centered radical (Eq. 35.1) and a radical initiator (Eq. 35.2)

from the reaction mechanism depicted in Scheme 37. In the Giese reaction, the tin hydride reduces the initial alkyl radical and the radical adduct at approximatively the same rate. Therefore, in order to favor the product of conjugate addition, it is compulsory to work with low concentration of tin hydride. In the catecholborane mediated reaction, the initial radical reacts much slower than the radical adduct with the PTOC-OMe chain transfer reagent. Indeed, a nucleophilic alkyl radical adds more slowly to the sulfur atom of a thiocarbonyl group than a radical having a marked electrophilic character such as the radical adduct.

B-alkylcatecholboranes, prepared by rhodium(I)-catalyzed hydroboration of alkenes, are suitable radical precursors for conjugate addition to activated olefins. This procedure is particularly useful for the control of the regio- and

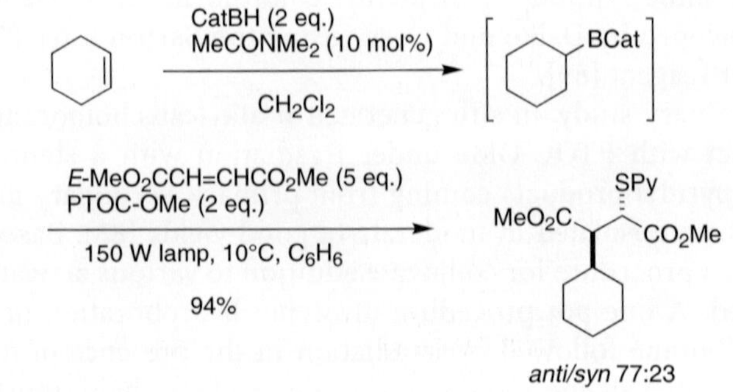

Scheme 36 PTOC-OMe mediated conjugate addition of *B*-alkylcatecholborane to activated alkenes

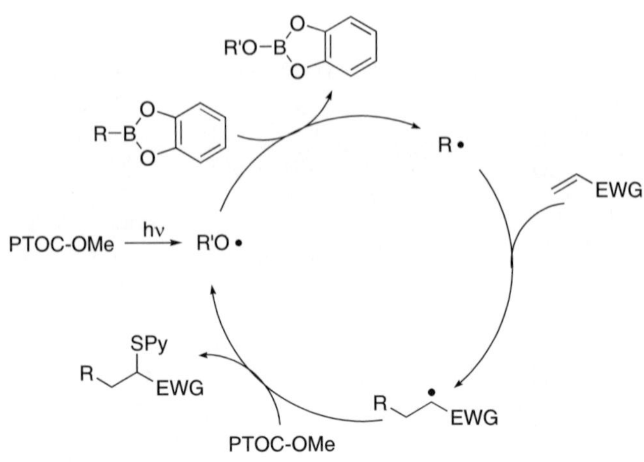

Scheme 37 Radical chain mechanism for the conjugate addition of *B*-alkylcatecholboranes to activated olefins (R = alkyl group; EWG = electron withdrawing group; R′O˙ =MeOC(O)O, MeO˙)

Brown proposed a mechanism where the enolate radical resulting from the radical addition reacts with the trialkylborane to give a boron enolate and a new alkyl radical that can propagate the chain (Scheme 24) [61]. The formation of the intermediate boron enolate was confirmed by ^{1}H NMR spectroscopy [66, 67]. The role of water present in the system is to hydrolyze the boron enolate and to prevent its degradation by undesired free-radical processes. This hydrolysis step is essential when alkynones [68] and acrylonitrile [58] are used as radical traps since the resulting allenes or keteneimines respectively, react readily with radical species. Maillard and Walton have shown by ^{11}B NMR, ^{1}H NMR und IR spectroscopy, that triethylborane does complex methyl vinyl ketone, acrolein and 3-methylbut-3-en-2-one. They proposed that the reaction of triethylborane with these traps involves complexation of the trap by the Lewis acidic borane prior to conjugate addition [69].

The reaction between trialkylboranes and enones has found some interesting synthetic applications. An example is the preparation of prostaglandin precursors from *exo*-methylene cyclopentanone, generated in situ from a Mannich base. After dehydrogenation, a second conjugate addition of trioctylborane was used to introduce the ω-chain (Scheme 25) [70].

Several attempts to take advantage of the intermediate boron enolate to achieve tandem conjugate addition-aldol reaction have been proposed [71]. Recently, Chandrasekhar [72] reported the addition of triethylborane to methyl vinyl ketone followed by the in situ trapping of the enolate by aromatic aldehyde (Scheme 26).

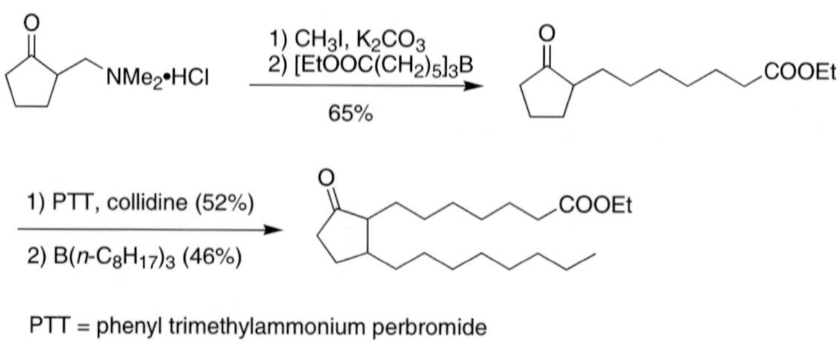

Scheme 25 Conjugate addition of a functionalized trialkylborane

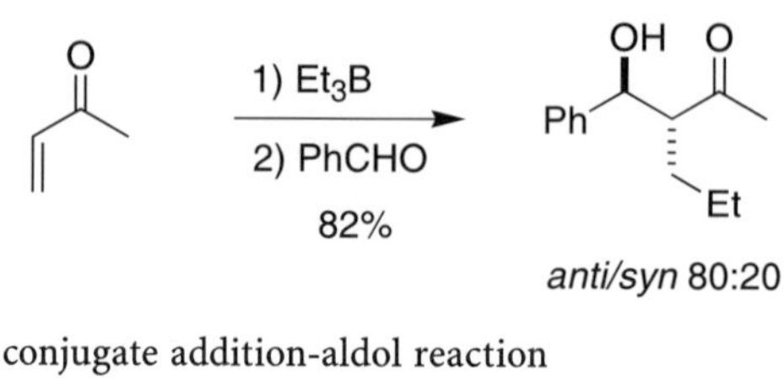

Scheme 26 Tandem conjugate addition-aldol reaction

chemoselectivity of such tandem processes [91]. A one-pot enantioselective hydroboration-radical conjugate addition was successfully performed. For example, the reaction between norbornene and methyl methacrylate as radical trap affords the product of conjugate addition in 68% yield and 85% ee (after

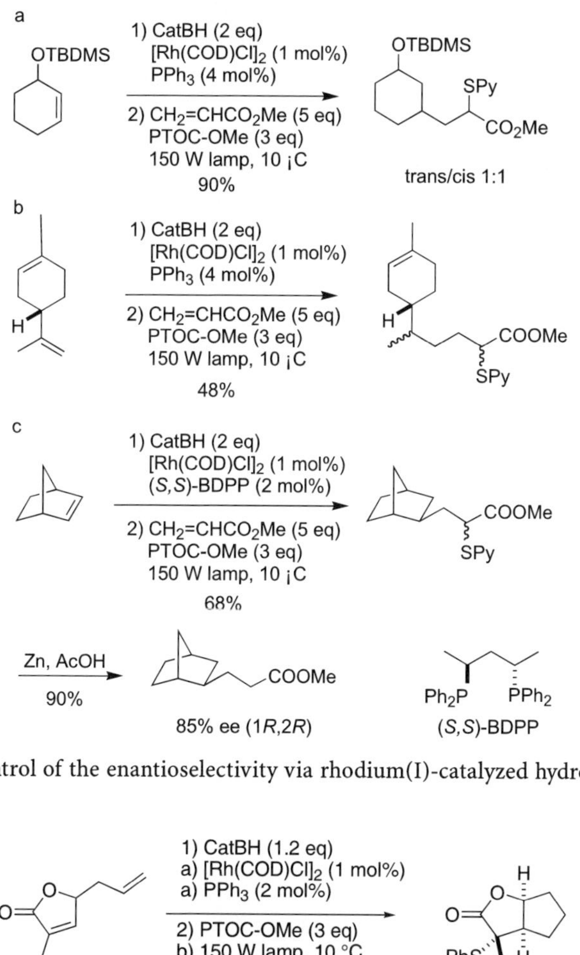

Scheme 38 Control of the enantioselectivity via rhodium(I)-catalyzed hydroboration

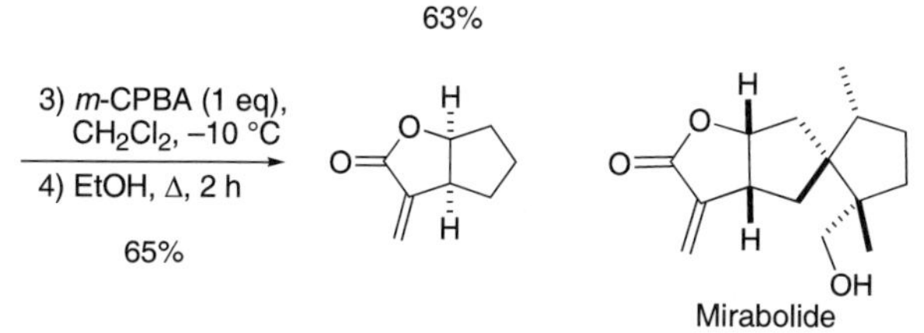

Scheme 39 Preparation of α-methylenelactone

desulfurization) using $[Rh(COD)Cl]_2$ and the chiral diphosphine (*S*,*S*)-BDPP as catalyst for the hydroboration step (Scheme 38).

The rhodium-catalyzed hydroboration has opened the way to cyclization reactions starting from dienes [92]. For instance, rhodium-catalyzed hydroboration of the terminal alkenyl group of an α,β-unsaturated lactone followed by reaction with the PTOC-OMe chain transfer reagent afforded the bicyclic α-*S*-pyridyl lactone in 63% yield (Scheme 39). After oxidation of the sulfide with *m*-CPBA, thermal elimination of the sulfoxide afforded the corresponding α-methylene lactone in 65% yield. Interestingly, such bicyclic α-methylenelactones are substructures that can be found in many natural products such as mirabolide [93].

3.3
Addition to Imines Derivatives

Intramolecular addition of trialkylboranes to imines and related compounds have been reported and the main results are part of review articles [94, 95]. Addition of ethyl radicals generated from Et_3B to aldimines affords the desired addition product in fair to good yield but low diaster control (Scheme 40, Eq. 40a) [96]. Similar reactions with aldoxime ethers [97], aldehyde hydrazones [97], and *N*-sulfonylaldimines [98] are reported. Radical addition to ketimines has been recently reported (Eq. 40b) [99]. Addition of triethylborane to 2H-azirine-3-carboxylate derivatives is reported [100]. Very recently, Somfai has extended this reaction to the addition of different alkyl radicals generated from trialkylboranes to a chiral ester of 2*H*-azirine-3-carboxylate under Lewis acid activation with CuCl (Eq. 40c) [101].

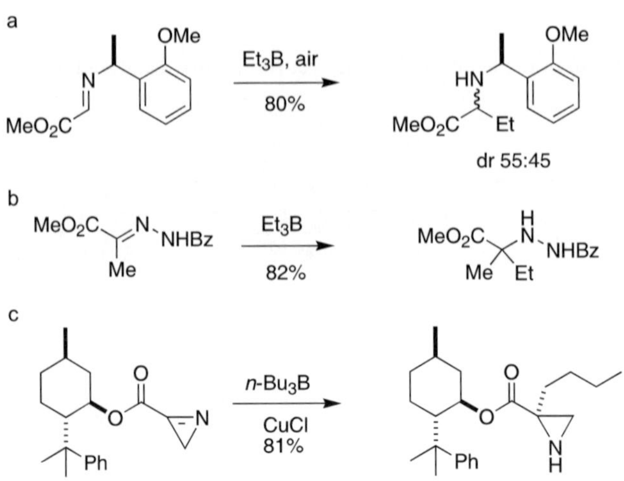

Scheme 40 Addition of trialkylboranes to aldimines, ketimines and related compounds

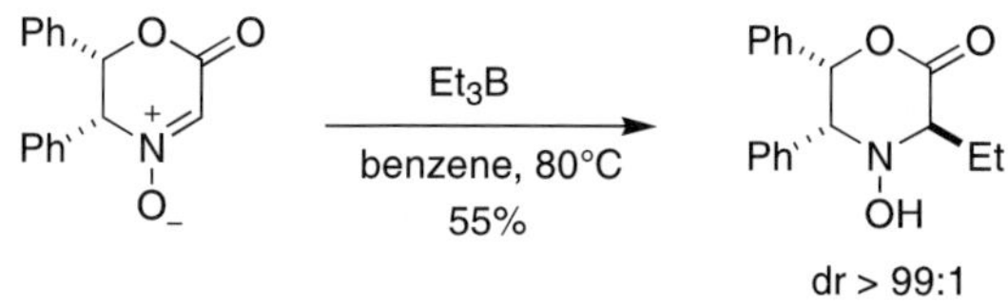

Scheme 41 Addition of triethylborane to nitrones

Naito reported a radical addition to nitrones that occurs with high stereocontrol (Scheme 41) [102].

Interestingly, most of the reactions reported with imines and related products such as oximes, oxime ethers, hydrazones and nitrones can be run in aqueous media [103].

3.4 C–C Bond Formation via *β*-Fragmentation Processes

Brown and Suzuki have shown that treatment of trialkylboranes with ethenyl- (Scheme 42, Eq. 42a) and ethynyloxiranes (Scheme 42, Eq. 42b) in the presence of a catalytic amount of oxygen, affords the corresponding allylic or allenic alcohols. The mechanism may involve the addition of alkyl radicals to the unsaturated system leading to 1-(oxiranyl)alkyl and 1-(oxiranyl)alkenyl radicals followed by rapid fragmentation to give alkoxyl radicals that finally complete the chain process by reacting with the trialkylborane [104–106].

The free radical substitution of *β*-nitrostyrene (2-nitroethenylbenzene) by trialkylboranes involves a radical addition to the *β*-position (*α* to the nitro group) followed by fragmentation of $NO_2^{\cdot}$ (Scheme 43). The reaction is *E*

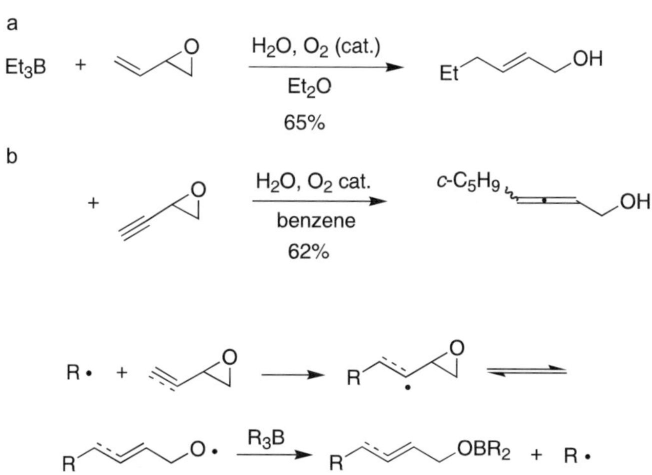

Scheme 42 Reaction of trialkylboranes with ethenyl- and ethynyloxiranes

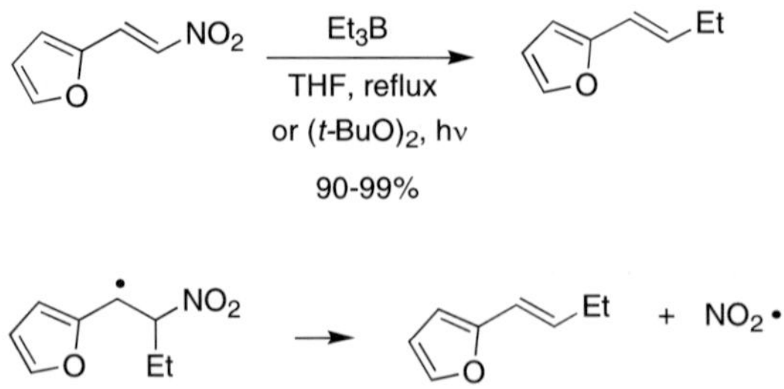

Scheme 43 Radical substitution of (2-nitrovinyl)arenes with Et_3B

selective and works with a broad range of trialkylboranes, allowing the introduction of tertiary, secondary and allylic carbon moieties [107].

Nozaki reported the reaction of trialkylboranes with styryl sulfoxides and sulfones. Alkyl radicals generated from trialkylboranes add at the β-position of β-styryl sulfoxides and sulfones (α- to the sulfur atom). The resulting radicals fragment and deliver the β-styryl adducts [108]. Interestingly, the sulfoxides eliminate very rapidly leading to partially stereospecific substitution (Scheme 44). The radical nature of the process is demonstrated by the presence of a side product derived from the solvent (THF) by hydrogen atom abstraction.

Radical allylation of *B*-alkylcatecholboranes using easily available allylsulfones has been described [109–111]. By using phenylsulfones, the fragmentation produces a stable phenylsulfonyl radical that reacts with *B*-alkyl-

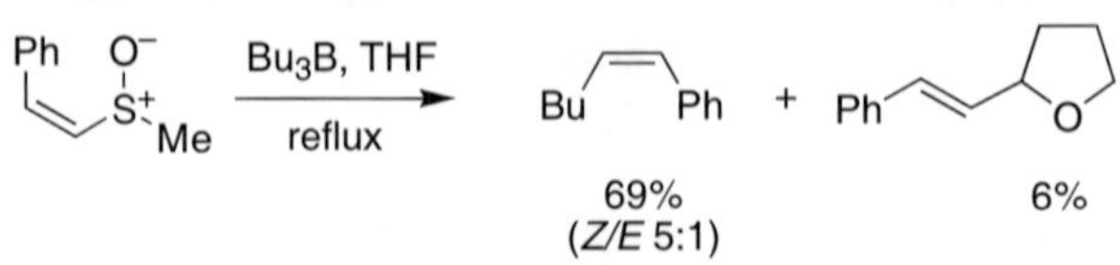

Scheme 44 Vinylation with styryl methyl sulfoxide

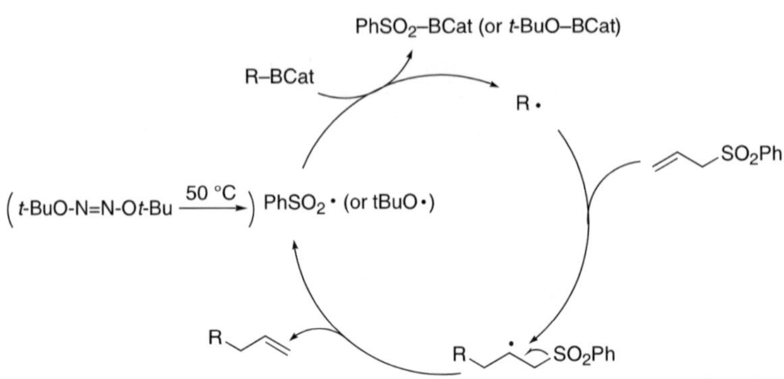

Scheme 45 Radical hydroallylation of alkenes

catecholborane to sustain the chain reaction (Scheme 45). Oxygen-centered radicals react efficiently with *B*-alkylcatecholboranes. Therefore, the easily available di-*tert*-butylhyponitrite was selected as an initiator due to its ability to furnish the *tert*-butoxyl radical at the refluxing temperature of dichloromethane. The thermal properties of this initiator allows one to run a one-pot hydroboration–radical reaction sequence by taking advantage of the very mild, efficient and cost effective hydroboration conditions developed by Fu [112].

The desired products were obtained in satisfactory to excellent yields by using only 1.2 equivalents of the allylsulfones with primary, secondary and tertiary alkyl radicals. Many different types of allylic sulfones bearing an ester group, a sulfonyl group, and a bromine atom react equally well (Scheme 46). The whole transformation represents formally a reductive allylation or hydroallylation of alkenes.

Interestingly, this allylation process seems to be very general. For instance, introduction of a dienyl moiety using penta-2,4-dienyl phenyl sulfone has been achieved (Scheme 47, Eq. 47a). The modest yield (50%) for the conversion is due to the instability of the dienyl sulfone which readily polymerizes. Finally, the radical nature of the process has been demonstrated by running

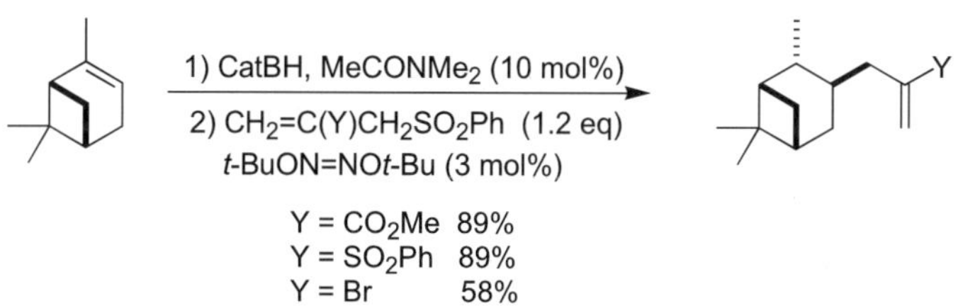

Scheme 46 Allylation of *B*-alkylcatecholboranes with allylsulfones

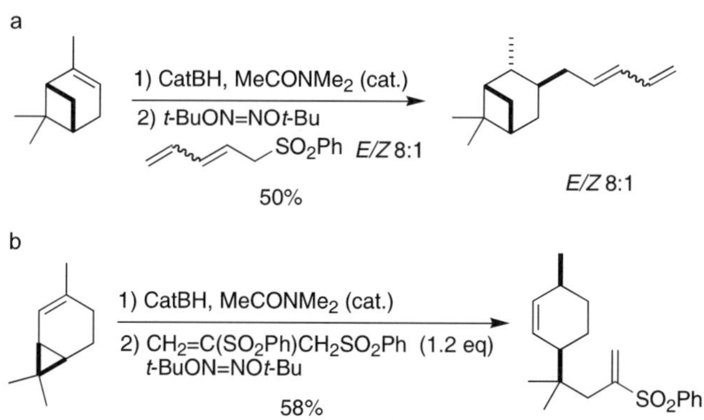

Scheme 47 Introduction of a dienyl moiety using penta-2,4-dienyl phenyl sulfone (Eq. 47.1) and hydroallylation of (+)-2-carene (Eq. 47.2)

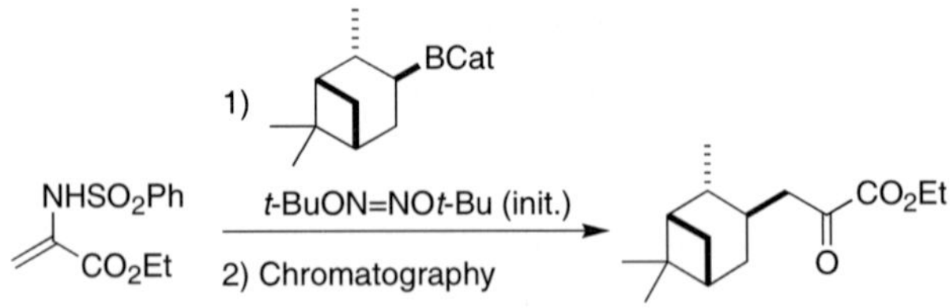

Scheme 48 Alkylation of ethyl pyruvate via reductive coupling of alkenes and ethyl 2-(benzenesulfonylamino)acrylate

an allylation reaction with (+)-2-carene (Eq. 47b). With this radical probe, the intermediate cyclopropylmethyl radical undergoes ring opening to a homoallylic radical that is trapped by the allylic sulfone to afford the corresponding monocyclic compound in 58% yield.

Radical coupling of *B*-alkylcatecholboranes, in situ generated from the corresponding alkenes, with ethyl 2-(benzenesulfonylamino)acrylate is reported (Scheme 48) [113]. This reaction represents an extension of the radical allylation of *B*-alkylcatecholboranes by allysulfones. This unique process allows one to prepare various α-ketoesters (alkylated pyruvates) in a straightforward manner. It also demonstrates the generality of the radical mediated C – C bond formation starting from organoboranes and allylic benzenesulfonyl derivatives.

4 Organoboranes as Chain Transfer Reagents

We have seen in the preceding chapters that trialkylboranes are useful radical initiators as well as efficient source of alkyl radicals. Organoboranes can also be used as chain-transfer reagents. This approach is used when the direct reaction between the radical precursor and the radical trap cannot proceed (Scheme 49, Eq. 49a). Alkyl radicals generated from the organoboranes are not involved in product formation, but they produce the radicals leading to products. For this purpose, an extra step such as an iodine atom transfer or a hydrogen abstraction is necessary. This point is schematically illustrated in Scheme 49 for a triethylborane mediated process (Eq. 49b). This reaction takes advantage of the high affinity of trialkylboranes for heteroatom centered radicals $X^{\cdot}$.

4.1 Via Iodine Atom Transfer

Radical addition of organoboranes to imines and related compounds is a promising alternative to the use of classical organometallic compound (see

a **Direct reaction**

$$\text{R–Y} + \text{A–X} \not\longrightarrow \text{R–A} + \text{Y–X}$$

$$\text{R}\bullet + \text{A–X} \longrightarrow \text{R–A} + \text{X}\bullet$$

$$\text{X}\bullet + \text{A–X} \not\longrightarrow \text{A}\bullet + \text{Et}_2\text{B–X}$$

b **Triethylborane mediated reaction**

$$\text{R–Y} + \text{A–X} + \text{Et}_3\text{B} \longrightarrow \text{R–A} + \text{Et–Y} + \text{Et}_2\text{B–X}$$

$$\text{X}\bullet + \text{Et}_3\text{B} \longrightarrow \text{Et}\bullet + \text{Et}_2\text{B–X}$$

$$\text{Et}\bullet + \text{R–Y} \longrightarrow \text{Et–Y} + \text{R}\bullet$$

$$\text{R}\bullet + \text{A–X} \longrightarrow \text{R–A} + \text{X}\bullet$$

R–Y = radical precursor
R–A = desired product
A–X = radical trap
X• = heteroatom centered radical
R• = radical involved in product formation

Scheme 49 Triethylborane as a chain transfer reagent for the conversion of R – Y to R – A

Sect. 3.2). However, this approach is limited to the few trialkylboranes that are easily available and cheap since only one of the three alkyl group is transferred. By using a triethylborane as a chain transfer reagent, the reaction could be extended to alkyl iodides as radical precursors. Bertrand [94, 114] and Naito [95, 97] reported both the use of triethylborane for the tin-free addition of alkyl iodides to imines. A typical example for a tentative of asymmetric addition to a glyoxylate imine is depicted in Scheme 50 (Eq. 50a). More recently additions to isatin imines were reported (Eq. 50b) as well as addition to 2*H*-aziridine-3-carboxylates by Lemos [100] and Somfai [101] (Eq. 50c).

Enantioselective radical addition to *N*-acyl hydrazone using triethylborane as chain transfer reagent has been reported by Friestad. Enantiomeric excesses up to 95% were obtained in the presence of copper(II)-bisoxazolines Lewis acid (Scheme 51) [115].

Tandem processes mediated by triethylborane involving conjugate addition to enones followed by aldol reaction are reported (Scheme 52, Eq. 52a). More recently, a tandem process involving addition of an isopropyl radical to an α,β-unsaturated oxime ether afforded an azaenolate intermediate that reacts with benzaldehyde in the presence of trimethylaluminum. The aldol product cyclizes to afford an isopropyl substituted γ-butyroloactone in 61% overall yield (Scheme 52) [116]. In these reactions, triethylborane is acting as a chain transfer reagent that delivers a boron enolate or azaenolate necessary for the aldolization process.

Alkenylation of alkyl iodides with β-nitrostyrene derivatives has been reported (Scheme 53) [117]. The reaction is, however, not strictly a chain pro-

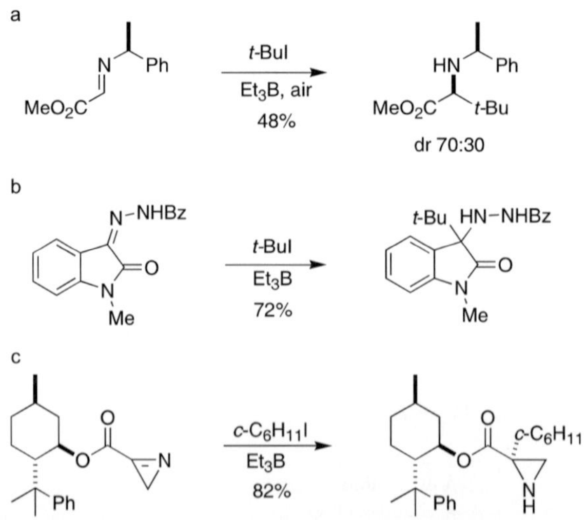

Scheme 50 Triethylborane mediated addition of alkyliodides to imines

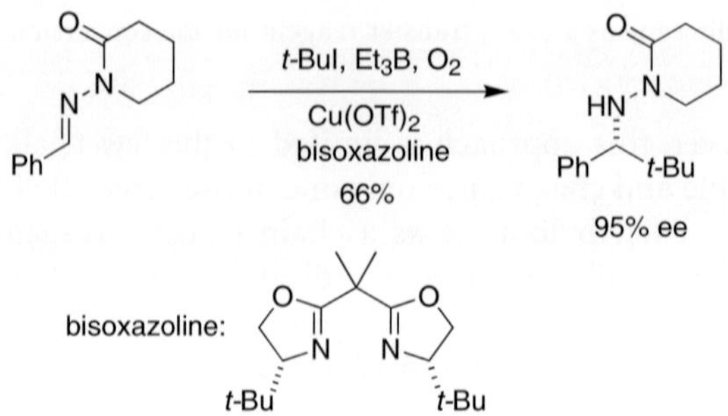

Scheme 51 Enantioselective addition to *N*-acyl hydrazones

cess since a stoichiometric amount of oxygen is necessary to run the reaction. The radical $NO_2^{\cdot}$ is presumably not sufficiently reactive towards triethylborane to sustain the chain process.

The radical carboazidation of alkenes has been achieved in water using triethylborane as initiator [118]. This efficient process is complete in one hour at room temperature in an open to air reaction vessel (Scheme 54, Eq. 54a). These new tin-free carboazidation conditions are environmentally friendly and allow to run reactions with an excess of either the alkene or the radical precursor. They are also suitable for simple radical azidation of alkyl iodides as well as for more complex cascade reactions involving annulation processes (Eq. 54b). In both reactions (Eq. 54a and 54b), an excess of triethylborane (3 equivalents) is required to obtain a good yield. This may be an indication that the chain process, more precisely the reaction between the phenylsulfonyl radical and Et_3B, is not efficient.

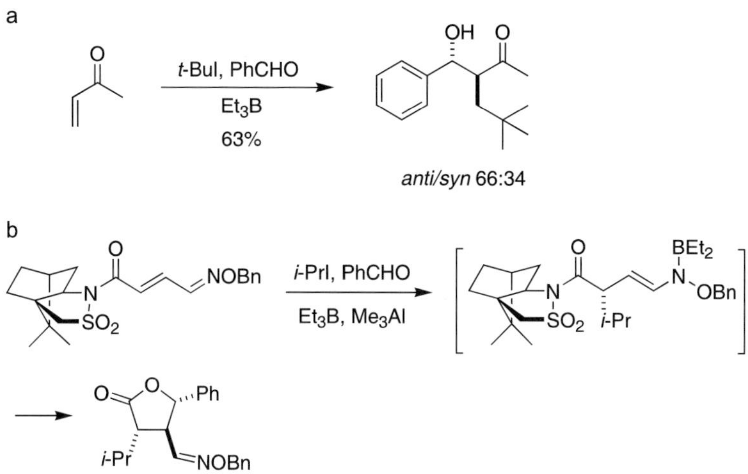

Scheme 52 Tandem radical addition aldol reaction

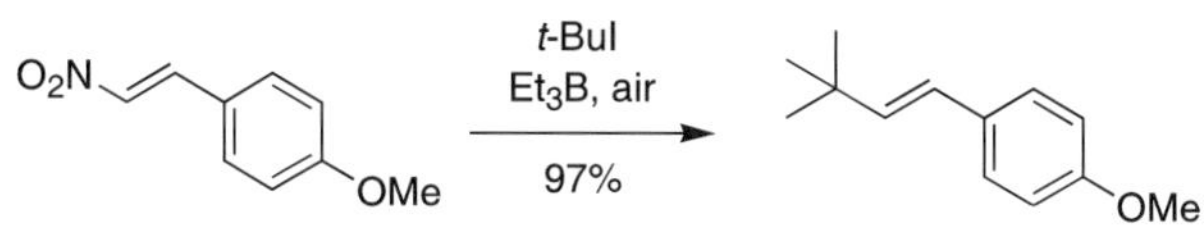

Scheme 53 Alkenylation of alkyl iodides

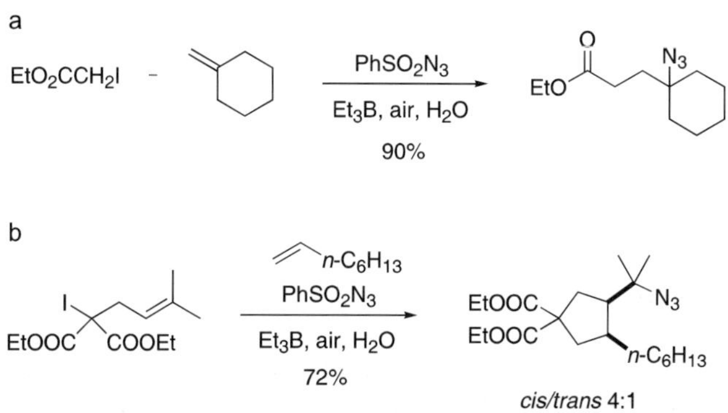

Scheme 54 Triethylborane mediated carboazidation

4.2 Via Hydrogen Atom Transfer

All the examples presented under Sect. 4.1 used an iodine atom transfer to generate the desired radicals. Another approach involving abstraction of hydrogen atom is also reported. For instance, ethers and acetals undergo direct intermolecular addition to aldehydes under treatment with Et_3B/air

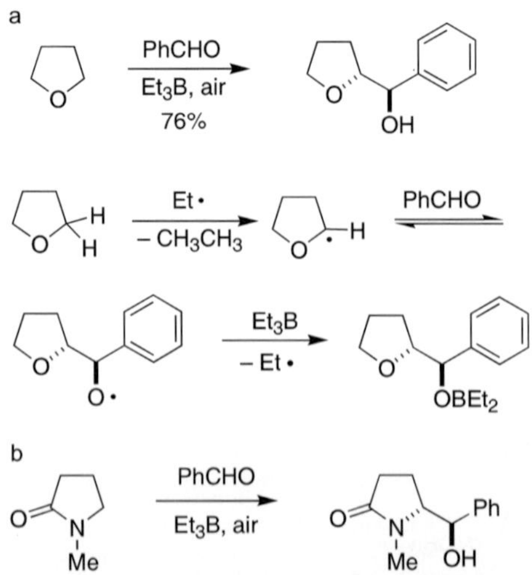

Scheme 55 Radical hydroxyalkylation of C – H bonds adjacent to oxygen and nitrogen

(Scheme 55, Eq. 55a) [119]. A plausible mechanism is depicted in Scheme 55 and involves radical addition of the 2-tetrahydrofuryl radical to the aldehyde followed by a rapid reaction of the alkoxyl radical with Et_3B. Triethylborane has a crucial role since by reacting with the alkoxyl radical it favors the formation of the condensation product relative to the β-fragmentation process (back reaction). A similar reaction with tertiary amines, amides and urea is also possible (Eq. 55b) [120].

5
Organoboron Compounds as Radical Reducing Agents

5.1
Complexes with Tertiary Amines

In pioneer work, Roberts investigated the use of amine-borane complexes as radical reducing agents. This research led him to develop the concept of polarity-reversal catalysis [121]. He found that a slow hydrogen atom abstraction step due to mismatched polarity can be replaced by two rapid steps with matched polarity. For example, the slow abstraction of an hydrogen atom from acetonitrile by a *tert*-butoxyl radical (mismatched polarity) is replaced by a rapid reduction of the *tert*-butoxyl radical with an amine-borane complex and by the abstraction of an hydrogen atom from acetonitrile by the amine-boryl radical [122, 123]. Attempts to use this concept for the kinetic resolution of chiral esters and lactones by using chiral amine-borane complexes lead to interesting enantioselectivities (Scheme 56) [124–126].

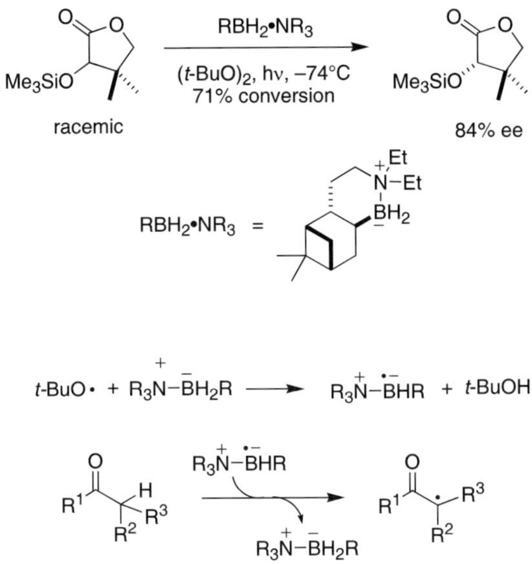

Scheme 56 Kinetic resolution of a γ-lactone

5.2 Complexes with Water and Alcohols

Wood [127] reported an innovative development of the Barton-McCombie deoxygenation of alcohols allowed to work under tin-free conditions. A trimethylborane-water complex proves to be an efficient reagent for the reduction of xanthates. Complexation of water by trimethylborane induces a strong decrease of O – H bond dissociation energy from 116 kcal/mol (water) to 86 kcal/mol (Me_3B-water complex).

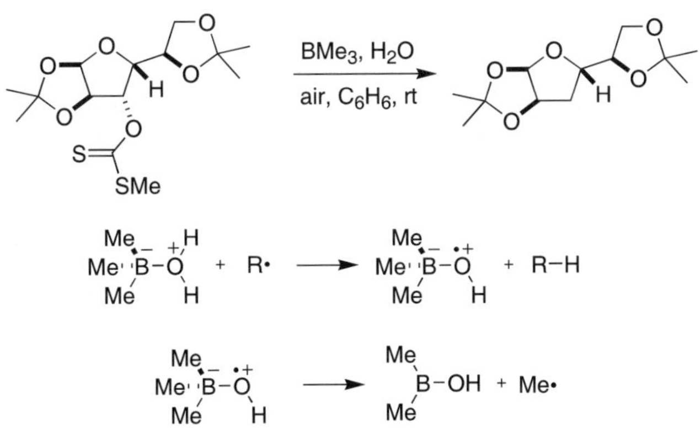

Scheme 57 Barton-McCombie deoxygenation with Me_3B-water complex

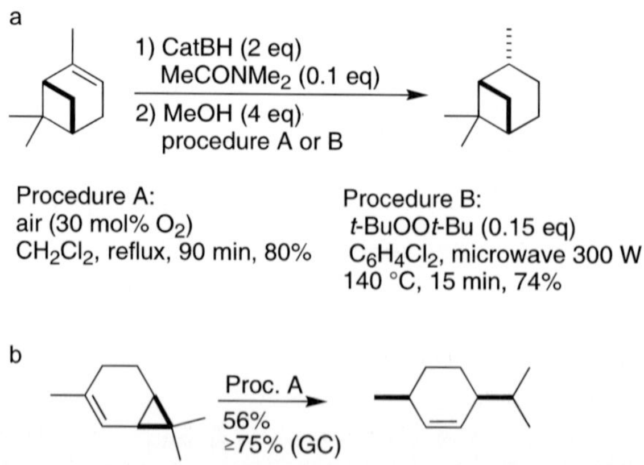

Scheme 58 Mild radical mediated reduction of organoborane with methanol

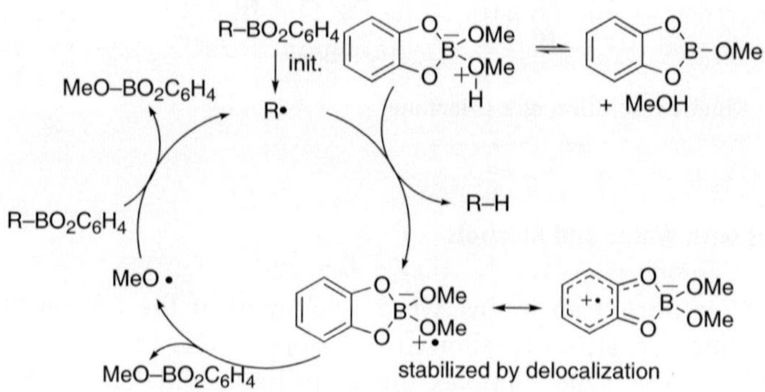

Scheme 59 Chain mechanism of the reduction of organoboranes

We made a similar observation when we reported a mild and efficient radical mediated reduction of organoboranes (Scheme 58, Eq. 58a) [128]. An in situ generated *B*-methoxycatecholborane-methanol complex acts as a reducing agent. The radical nature of the process was demonstrated by using (+)-2-carene as a radical probe (Eq. 58b). Water, ethanol and trifluoroethanol can be used instead of MeOH with very similar efficiency.

The reaction mechanism of this transformation is depicted in Scheme 59 and involves activation of the O – H bond of methanol by complexation with *B*-methoxycatecholborane. Interestingly, the reduction leads after fragmentation of the radical-ate complex to a methoxyl radical that reacts very efficiently with the *B*-alkylcatecholborane warranting an efficient chain process.

6
Conclusions

The tremendous development of the use of radicals in organic synthesis and the necessity of avoiding the use of tin derivatives because of its toxicity has led to a revival of the radical chemistry of organoboranes. The use of triethylborane as an initiator for radical chain reactions is now part of the classical arsenal of organic chemists. Generation of more complex and functionalized radicals from organoboranes is of great interest since it allows one to consider olefins as a potential source of radicals. So far, the generation of radicals has not been extended to alkenyl and aryl radicals, but rapid progress is expected in this field. Interestingly, organoboranes could also play the role of chain transfer reagents in radical processes. Due to the particularly rich reactivity of boron derivatives, the design of tandem processes involving radical and non-radical reactions is now possible. Finally, boron derivatives are promising reagents for activating water and alcohols and making them suitable reagents for the reduction of radicals. Spectacular development in this particular field is expected in the near future.

References

1. Brown HC, Midland MM (1972) Angew Chem Int Ed Engl 11:692
2. Ghosez A, Giese B, Zipse H (1989) (Houben-Weyl) E19a:753
3. Ollivier C, Renaud P (2001) Chem Rev 101:3415
4. Davies AG, Roberts BP (1972) Acc Chem Res 5:387
5. Davies AG (1974) Pure Appl Chem 39:497
6. Nozaki K, Oshima K, Utimoto K (1987) J Am Chem Soc 109:2547
7. Yorimitsu H, Oshima K (2001) In: Renaud P, Sibi MP (eds) Radicals in Organic Synthesis. Wiley, Weinheim 1:11
8. Batey RA, Smil DV (1999) Angew Chem Int Ed 38:1798
9. Schaffner A-P, Renaud P (2004) Eur J Org Chem :2291
10. Miura K, Ichinose Y, Nozaki K, Fugami K, Oshima K, Utimoto K (1989) Bull Chem Soc Jpn 62:143
11. Evans PA, Roseman JD (1996) J Org Chem 61:2252
12. Salazar KL, Nicholas KM (2000) Tetrahedron 56:2211
13. Yorimitsu H, Shinokubo H, Oshima K (2002) Synlett :674
14. Yamazaki O, Togo H, Nogami G, Yokoyama M (1997) Bull Chem Soc Jpn 70:2519
15. Gonzalez-Lopez de Turiso F, Curran DP (2005) Org Lett 7:151
16. Nakamura T, Yorimitsu H, Shinokubo H, Oshima K (1999) Synlett :1415
17. Barton DHR, Parekh SI, Tse C-L (1993) Tetrahedron Lett 34:2733
18. Yorimitsu H, Shinokubo H, Oshima K (2000) Chem Lett :104
19. Lee (2002) Tetrahedron Lett 43:7295
20. Miura K, Oshima K, Utimoto K (1993) Bull Chem Soc Jpn 66:2356
21. Miura K, Oshima K, Utimoto K (1993) Bull Chem Soc Jpn 66:2348
22. Ichinose Y, Nozaki K, Wakamatsu K, Oshima K, Utimoto K (1987) Tetrahedron Lett 28:3709

23. Ichinose Y, Wakamatsu K, Nozaki K, Birbaum J-L, Oshima K, Utimoto K (1987) Chem Lett :1647
24. Tanaka S, Nakamura T, Yorimitsu H, Shinokubo H, Oshima K (2000) Org Lett 2:1911
25. Taniguchi M, Oshima K, Utimoto K (1993) Chem Lett :1751
26. Usugi S, Yorimitsu H, Oshima K (2001) Tetrahedron Lett 42:4535
27. Deprele S, Montchamp J-L (2001) J Org Chem 66:6745
28. Gautier A, Garipova G, Dubert O, Oulyadia H, Piettre SR (2001) Tetrahedron Lett 42:5673
29. Jessop CM, Parsons AF, Routledge A, Irvine DJ (2003) Tetrahedron: Asymmetry 14:2849
30. Takami K, Yorimitsu H, Oshima K (2004) Org Lett 6:4555
31. Hirano K, Fujita K, Shinokubo H, Oshima K (2004) Org Lett 6:593
32. Miura K, Tanigushi M, Nozaki K, Oshima K, Utimoto K (1990) Tetrahedron Lett 31:6391
33. Curran DP, Chen M-H, Spetzler E, Seong CM, Chang C-T (1989) J Am Chem Soc 111:8872
34. Byers J (2001) In: Renaud P, Sibi MP (eds) Radicals in Organic Synthesis. Wiley, Weinheim 1:72
35. Takeyama Y, Ichinose Y, Oshima K, Utimoto K (1989) Tetrahedron Lett 30:3159
36. Baciocchi E, Muraglia E (1994) Tetrahedron Lett 35:2763
37. Tang Y, Li C (2004) Org Lett 6:3229
38. Ichinose Y, Matsunaga S-I, Fugami K, Oshima K, Utimoto K (1989) Tetrahedron Lett 30:3155
39. Yorimitsu H, Nakamura T, Shinokubo H, Oshima K (1998) J Org Chem 63:8604
40. Yorimitsu H, Nakamura T, Shinokubo H, Oshima K, Omoto K, Fujimoto H (2000) J Am Chem Soc 122:11041
41. Nakamura T, Yorimitsu H, Shinokubo H, Oshima K (1998) Synlett :1351
42. Ikeda M, Teranishi H, Iwamura N, Ishibashi H (1997) Heterocycles 45:863
43. Ikeda M, Teranishi H, Nozaki K, Ishibashi H (1998) J Chem Soc, Perkin Trans 1:1691
44. Albrecht U, Wartchow R, Hoffmann HMR (1992) Angew Chem Int Ed 31:910
45. Woltering TJ, Hoffmann HMR (1995) Tetrahedron 51:7389
46. Kitagawa O, Yamada Y, Fujiwara H, Taguchi T (2001) Ang Chem Int Ed 40:3865
47. Sugimoto J, Miura K, Oshima K, Utimoto K (1991) Chem Lett :1319
48. Iseki K, Nagai Y, Kobayashi Y (1993) Tetrahedron Lett 34:2169
49. Iseki K, Nagai Y, Kobayashi Y (1994) Tetrahedron: Asym 5:961
50. Iseki K, Asada D, Takahashi M, Nagai T, Kobayashi Y (1996) Chem Pharm Bull 44:1314
51. Iseki K, Asada D, Takahashi M, Nagai T, Kobayashi Y (1994) Tetrahedron Lett 35:7399
52. Itoh Y, Mikami K (2005) Organic Letters 7:4883
53. Kita Y, Sano A, Yamaguchi T, Oka M, Gotanda K, Matsugi M (1997) Tetrahedron Lett 38:3549
54. Mero CL, Porter NA (1999) J Am Chem Soc 121:5155
55. Yorimitsu H, Oshima K (2002) Bull Chem Soc Jpn 75
56. Tsuritani T, Shinokubo H, Oshima K (2001) Org Lett 3:2709
57. Montermini F, Lacote E, Malacria M (2004) Org Lett 6:921
58. Suzuki A, Arase A, Matsumoto H, Itoh M, Brown HC, Rogic MM, Rathke MW (1967) J Am Chem Soc 89:5709
59. Brown HC, Rogic MM, Rathke MW, Kabalka GW (1967) J Am Chem Soc 89:5709
60. Brown HC, Kabalka GW, Rathke MW, Rogic MM (1968) J Am Chem Soc 90:4165

61. Kabalka GW, Brown HC, Suzuki A, Honma S, Arase A, Itoh M (1970) J Am Chem Soc 92:710
62. Brown HC, Kabalka GW (1970) J Am Chem Soc 92:712
63. Brown HC, Kabalka GW (1970) J Am Chem Soc 92:714
64. Suzuki A, Nozawa S, Itoh M, Brown HC, Negishi E, Gupta SK (1969) J Chem Soc, Chem Commun :1009
65. Brown HC, Negishi E (1971) J Am Chem Soc 93:3777
66. Fenzl W, Köster R, Zimmermann H-J (1975) Liebigs Ann Chem :2201
67. Mukaiyama T, Inomata K, Muraki M (1973) J Am Chem Soc 95:967
68. Suzuki A, Nozawa S, Itoh M, Brown HC, Kabalka GW, Holland GW (1970) J Am Chem Soc 92:3503
69. Beraud V, Gnanou Y, Walton JC, Maillard B (2000) Tetrahedron Lett 41:1195
70. Attanasi O, Baccolini G, Caglioti l, Rosini G (1973) Gazz Chim Ital 103:31
71. Nozaki K, Oshima K, Utimoto K (1991) Bull Chem Soc Jpn 64:403
72. Chandrasekhar S, Narsihmulu C, Reddy NR, Reddy MS (2003) Tetrahedron Lett 44:2583
73. Toru T, Watanabe Y, Mase N, Tsusaka M, Hayakawa T, Ueno Y (1996) Pure Appl Chem 68:711
74. Toru T, Watanabe Y, Tsusaka M, Ueno Y (1993) J Am Chem Soc 115:10464
75. Mase N, Watanabe Y, Toru T (1998) J Org Chem 63:3899
76. Mase N, Watanabe Y, Ueno Y, Toru T (1997) J Org Chem 62:7794
77. Mase N, Watanabe Y, Ueno Y, Toru T (1998) J Chem Soc, Perkin Trans 1:1613
78. Ollivier C, Renaud P (1999) Chem Eur J 5:1468
79. Baban JA, Goodchild NJ, Roberts BP (1986) J Chem Soc, Perkin Trans 2:157
80. Forster AH (1999) PhD Thesis, Université de Fribourg, Switzerland, Diss Nr 1242
81. Hawthorne MF, Reintjes M (1964) J Am Chem Soc 86:951
82. Hawthorne MF, Reintjes M (1965) J Am Chem Soc 87:4585
83. Kabalka GW (1971) J Organomet Chem 33:C25
84. Bieber LW, Rolim Neto PJ, Generino RM (1999) Tetrahedron Lett 40:4473
85. Kumli E, Renaud P, submitted for publication
86. Newcomb M, Kumar MU, Boivin J, Crépon E, Zard SZ (1991) Tetrahedron Lett :45
87. Beckwith ALJ, Davison IGE (1991) Tetrahedron Lett 32:49
88. Ollivier C, Renaud P (2000) Angew Chem Int Ed 39:925
89. Cadot C, Cossy J, Dalko PI (2000) Chem Commun :1017
90. Giese B (1983) Angew Chem Int Ed 23:753
91. Renaud P, Ollivier C, Weber V (2003) J Org Chem 68:5769
92. Becattini B, Ollivier C, Renaud P (2003) Synlett :1485
93. Bohlmann F, Ludwig GW, Jakupovic J, King RM, Robinson H (1984) Liebigs Ann Chem :228
94. Bertrand M, Feray L, Gastaldi S (2002) CR Acad Sc Chimie 5:623
95. Miyabe H, Ueda M, Naito T (2004) Synlett 7:1140
96. Bertrand MP, Coantic S, Feray L, Nouguier R, Perfetti P (2000) Tetrahedron 56:3951
97. Miyabe H, Shibata R, Sangawa M, Ushiro C, Naito T (1998) Tetrahedron 54:11431
98. Miyabe H, Ueda M, Naito T (2000) Chem Commun :2059
99. Miyabe H, Yamaoka Y, Takemoto Y (2005) J Org Chem 70:3324
100. Alves MJ, Fortes G, Guimaraes E, Lemos A (2003) Synlett :1403
101. Risberg E, Fischer A, Somfai P (2005) Tetrahedron 61:8443
102. Ueda M, Miyabe H, Teramachi M, Miyata O, Naito T (2003) Chem Commun :426
103. Miyabe H, Ueda M, Naito T (2000) J Org Chem 65:5043

104. Suzuki A, Miyaura N, Itoh M, Brown HC, Holland GW, Negishi E (1971) J Am Chem Soc 93:2792
105. Suzuki A, Miyaura N, Itoh M, Brown HC, Jacob III P (1973) Synthesis :305
106. Miyaura N, Itoh M, Sasaki N, Suzuki A (1975) Synthesis :317
107. Yao C-F, Chu C-M, Liu J-T (1998) J Org Chem 63:719
108. Miyamoto N, Fukuoka D, Utimoto K, Nozaki H (1974) Bull Chem Soc Jpn 47:503
109. Schaffner A-P, Renaud P (2003) Angew Chem Int Ed 42:2658
110. Schaffner A-P, Becattini B, Ollivier C, Weber V, Renaud P (2003) Synthesis :2740
111. Darmency V, Scanlan EM, Schaffner A-P, Renaud P (2005) Organic Syntheses, in press
112. Garett CE, Fu GC (1996) J Org Chem 61:3224
113. Darmency V, Renaud P (2005) Chimia 59:109
114. Bertrand MP, Feray L, Nouguier R, Stella L (1998) Synlett :780
115. Friestad GK, Shen Y, Ruggles EL (2003) Angew Chem Int Ed 42:5061
116. Ueda M, Miyabe H, Sugino H, Miyata O, Naito T (2005) Angew Chem Int Ed 44:6190
117. Liu J-T, Jang Y-J, Shih Y-K, Hu S-R, Chu C-M, Yao C-F (2001) J Org Chem 66:6021
118. Panchaud P, Renaud P (2004) J Org Chem 69:3205
119. Yoshimitsu T, Arano Y, Nagaoka H (2005) J Org Chem 70:2342
120. Yoshimitsu T, Arano Y, Nagaoka H (2005) J Am Chem Soc 127:11610
121. Roberts BP (1999) Chem Soc Rev 28:25
122. Paul V, Roberts BP (1987) Chem Commun :1322
123. Paul V, Roberts BP (1988) J Chem Soc Perkin Trans 2:1183
124. Dang H-S, Diart V, Roberts BP (1994) J Chem Soc Perkin Trans 1:1033
125. Mok PLH, Roberts BP, McKetty PT (1993) J Chem Soc Perkin Trans 2:665
126. Dang H-S, Diart V, Roberts BP, Tocher DA (1994) J Chem Soc Perkin Trans 2:1039
127. Spiegel DA, Wiberg KB, Schacherer LN, Medeiros MR, Wood JL (2005) J Am Chem Soc 127:12513
128. Pozzi D, Scanlan EM, Renaud P (2005) J Am Chem Soc 127:14204

Top Curr Chem (2006) 263: 107–162
DOI 10.1007/128_027

Published online: 19 January 2006

Enantioselective Radical Reactions

Jake Zimmerman · Mukund P. Sibi (✉)

North Dakota State University, Fargo, North Dakota, USA
mukund.sibi@ndsu.edu

Abstract Over the past two decades, many researchers have been interested in "taming" the reactive free radical intermediate and utilizing it in enantioselective transformations. This review highlights the substantial progress made in this area of synthetic organic chemistry. The main classes of radical reactions that have received the most attention include atom transfer, conjugate addition, addition and trapping, and electron transfer reactions. Also, the possibility of establishing multiple stereocenters in one simple transformation makes enantioselective free radical chemistry a very attractive tool for organic chemists.

Keywords Atom transfer · Chiral Lewis acid · Conjugate addition · Enantioselective radical reactions · Electron transfer

1 Introduction

Radical chemistry has seen tremendous progress in the past two decades and can now be considered as an eminent sub discipline in synthetic organic chemistry [1–6]. Diastereoselective radical chemistry is well established and many examples of enantioselective radical reactions have appeared in the recent literature. For reviews on diastereoselective radical chemistry see [7–11]; for reviews on enantioselective radical chemistry see [12–16]; and for reviews on conjugate additions, see [17, 18]. This review will detail different ways to introduce asymmetry during a radical reaction. These transformations can be broadly classified into atom transfer reactions, reductive alkylations, fragmentations, addition and trapping experiments, and electron transfer reactions.

2 Atom/Group Transfer Reactions

Atom/group transfer reactions can be broadly defined as those that involve the transfer of an atom (or a group) from a chain transfer agent to a radical species to generate another radical in a potentially chain propagating step. Two major classes of atom transfer reactions are the transfer of hydrogen or a halogen atom. Although, many examples of group transfer reactions are known, enantioselective examples are missing at present.

2.1 Hydrogen Atom Transfer

Hydrogen atom transfer implies the transfer of hydrogen atoms from the chain carrier, which is the stereo-determining step in enantioselective hydrogen atom transfer reactions. These reactions are often employed as a functional group interconversion step in the synthesis of many natural products wherein an alkyl iodide or alkyl bromide is converted into an alkane, which, in simple terms, is defined as reduction [19, 20]. Most of these reactions can be classified as diastereoselective in that the selectivity arises from the substrate. Enantioselective H-atom transfer reactions can be performed in two distinct ways: (1) by H-atom transfer from an achiral reductant to a radical complexed to a chiral source or alternatively; (2) by H-atom transfer from a chiral reductant to a radical.

2.1.1 Chiral Lewis Acid

Chiral Lewis acid mediated reductions at carbon atom α to a carbonyl group can be carried out either by generation of the radical from α-halo carbonyl compounds or from conjugate addition to a β-carbon atom. Scheme 1 shows both a C-centered and an enol form of a radical generated in the presence of a chiral Lewis acid. The hydrogen atom can be delivered selectively to one face of either **2** or **3**. Murakata et al. described the reduction of α-alkyl-α-iododihydrocoumarins using stoichiometric amounts of MgI_2 and a C_2-symmetric adiamine as a chiral Lewis acid in the presence of Bu_3SnH as a hydrogen atom source (Scheme 2) [21].

The results from reduction of the α-alkyl-α-iododihydrocoumarins **4a–d** are shown in Scheme 2. It was found that the substrate concentration greatly affected the observed enantioselectivities (compare entries 1 and 2). This may suggest that under dilute conditions there is a higher amount of un-

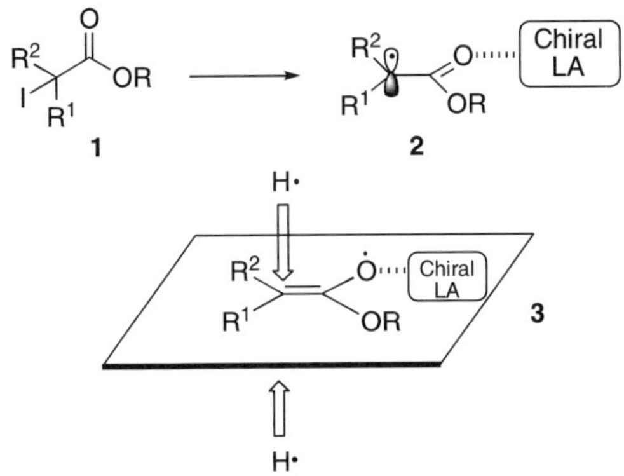

Scheme 1 Chiral Lewis acid controlled H-atom transfer

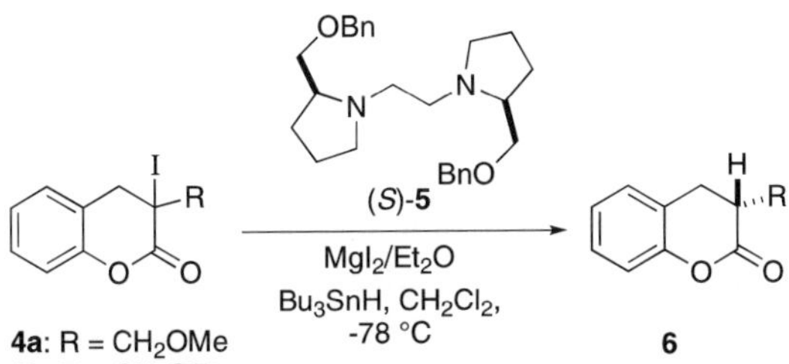

4a: R = CH_2OMe
4b: R = CH_2OEt
4c: R = CH_2OBn
4d: R = Me

Entry	Substrate	Conc. of **4** (mM)	Yield (%)	ee (%)
1	**4a**	11	75	12 (*R*)
2	**4a**	36	88	62 (*R*)
3	**4b**	37	84	65 (*R*)
4	**4c**	38	89	58 (*R*)
5	**4d**	35	78	30 (*S*)

Scheme 2 Reduction of α-iodo lactones

complexed enol leading to product with low enantioselectivity. Under higher concentration (36 mM **4a**), however, excellent chemical yields and good to moderate selectivities were achieved. The reaction efficiency depends on the rate of H-atom transfer: reactions using Ph_3SnH were much slower than with Bu_3SnH (5 h vs. 40 min) but chemical yields and selectivities were similar. Tris(trimethylsilyl)silane (TTMSS), a weaker H-atom donor, proved ineffective and no reaction was observed. The nature of the binding of substrate to the chiral Lewis acid is not apparent. Substrates **4a–c** give higher selectivities than **4d** and leads us to conclude that bidentate binding of the substrates is essential for higher ee's in this system.

In one of the earliest reports on enantioselective radical reactions, chiral Lewis acid mediated conjugate addition followed by enantioselective H-atom transfer α to a carbonyl was reported by Sato and co-workers (Scheme 3) [22]. The single point binding chiral aluminum complex presumably coordinates to the carbonyl oxygen of the lactone as shown in **10**. The strong Lewis acidity of the aluminum complex activates the substrate **7** to nucleophilic conjugate addition, which is followed by an enantioselective H-atom transfer from Bu_3SnH in a chiral environment provided by BINOL ligand in **8**. Only 28% ee was observed for product **9**.

The sulfonyl group is achiral in itself but has prochiral oxygens, and sultams have served as important auxiliaries in chiral reactions. Scheme 4 shows a simple example of selective coordination of a chiral Lewis acid to the sulfonyl oxygen.

A highly enantioselective hydrogen atom transfer reaction to α-methacrylates using a novel naphthosultam template has been reported recently [23]. In conjugate radical additions, the high enantioselectivity is attributed to the control of various rotamers of the starting material with reaction occurring from one reactive conformation. For α-alkyl-substituted systems, favorable conformation cannot be easily predicted, and it is possible for an *s-cis*, *s-trans*, or an alternate twisted conformer to predominate (Scheme 5). Previous results show that twisting does occur, and the relief of steric strain in the substrate overcomes any additional stabilization obtained by π conjugation. In the chiral Lewis acid complexed system, hydrogen atom

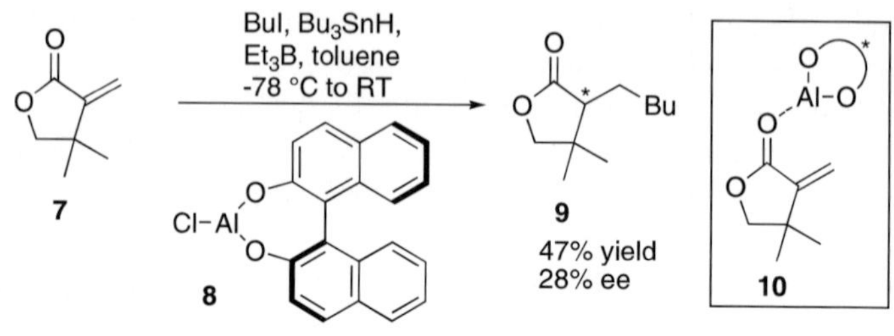

Scheme 3 Reductions mediated by aluminum-BINOL

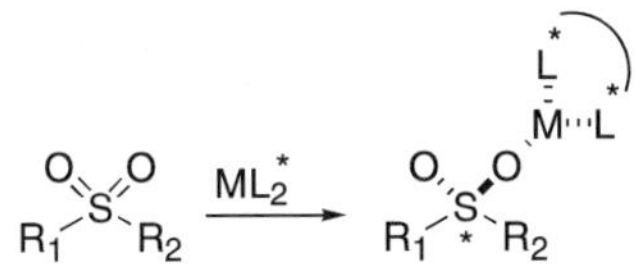

Scheme 4 Asymmetric induction via the sulfonyl group

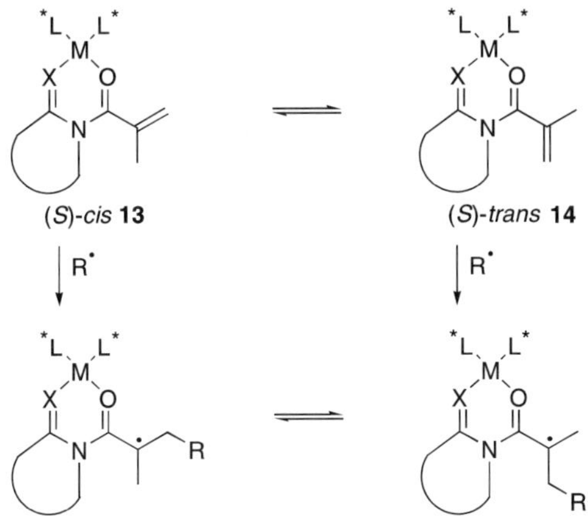

Scheme 5 Rotamer control

transfer to the intermediate radical should occur selectively from one reactive conformation after conjugate radical addition. If hydrogen atom transfer can occur much faster than conformational interconversion (**15–16**), then maximum face shielding from a substrate-chiral Lewis acid complex should provide high selectivity.

It was found that achiral template 1,8-naphthosultam in combination with C_2-symmetric bis-oxazoline ligand **19** efficiently controlled the rotamer of the acyl side chain of α-methacrylates. Scheme 6 shows the results of nucleophilic conjugate radical additions followed by hydrogen atom transfer to these substrates. In the absence of any Lewis acid, no reaction occurred. In the presence of 1 equivalent of $MgBr_2 \cdot Et_2O$ and chiral ligand **19**, high yield and selectivity was obtained on addition of isopropyl radical (entry 2). Substoichiometric amounts of chiral Lewis acid gave similar results (entry 3). When Ph_3SnH, a slow hydrogen atom donor, was used, the same selectivity was essentially obtained; however, the chemical yield was greatly reduced (entry 4). Other radicals were also successfully added, and methoxymethyl radical furnished the best selectivity (entries 5 and 6).

Radical reactions are known to proceed by an early transition state, which structurally resembles the starting complex. Here, it is assumed that the hydrogen atom transfer occurs rapidly as compared to any rotamer interconver-

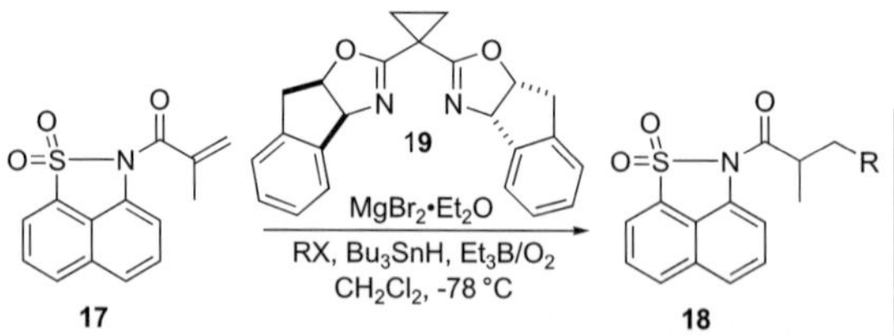

Entry	R	Lewis Acid (equiv.)	Yield (%)	ee (%)
1	*i*-Pr	none	<5	0
2	*i*-Pr	$MgBr_2 \cdot Et_2O$ (1.0)	90	78
3	*i*-Pr	$MgBr_2 \cdot Et_2O$ (0.3)	80	80
4[a]	*i*-Pr	$MgBr_2 \cdot Et_2O$ (0.3)	55	80
5	*t*-Bu	$MgBr_2 \cdot Et_2O$ (0.3)	84	89
6	$MeOCH_2$	$MgBr_2 \cdot Et_2O$ (0.3)	89	90

[a] Ph_3SnH was used

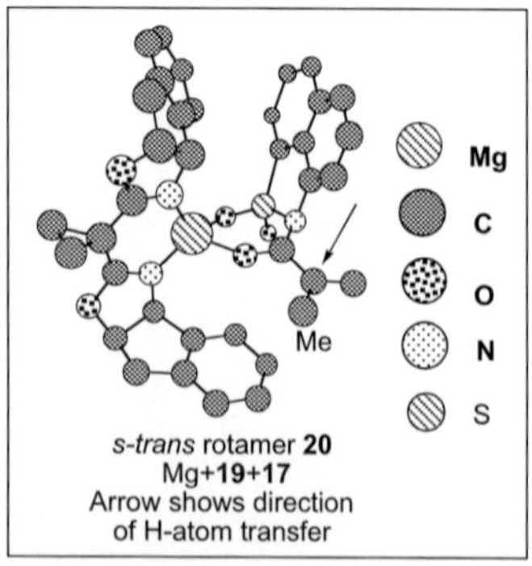

Scheme 6 Sultam templates in enantioselective H-atom transfer

sion, and thus, the precursor geometry impacts product stereochemistry. The absolute configuration of the isopropyl radical addition product was determined to be (*S*). Product stereochemistry analysis suggests that the reaction should occur from the conformer shown in Scheme 6 (**20**) with hydrogen atom transfer taking place from the *re*-face of the radical intermediate [23]. The reasons for the preference for this rotamer are not well understood.

Toru and co-workers have recently reported another example of asymmetric induction through the sulfonyl group [24]. Conjugate addition of nucleophilic alkyl radicals followed by enantioselective hydrogen atom transfer to α-sulfonyl radicals with a benzimidazolyl template were studied (Scheme 7). They first examined hydrogen atom transfer to an α-sulfonyl radical generated from the addition of a *t*-butyl radical to 2-propenyl sulfones (R = Me) **21** (entries 1 and 2). Various Lewis acids were screened and $Zn(OTf)_2$ with phenyl bis-oxazoline ligand **23** gave the best yield and selectivity. A screening of hydrogen atom donors showed that trimethyltin hydride (Me_3SnH), a slightly less hindered hydride, gave similar results as Bu_3SnH, albeit with slightly lower selectivity. This suggests that the stereoselectivity depends on the sterics of the tin reagents.

Addition of *t*-butyl radical was unsuccessful when 1-phenylethenyl sulfones (R = Ph) were used, yet the less sterically demanding ethyl radical addition proceeded smoothly under the standard radical conditions (entries 3 and 4). Interestingly, when bulky hydrogen atom donors, such as Ph_3SnH and TTMS were used, longer reaction times were required, but the selectivity was the same as that obtained with Bu_3SnH.

The absolute stereochemistry of the products formed from radical addition onto 2-propenyl sulfone and 1-phenylethenyl sulfone were found to be opposite. The reactions proceed through a five-membered transition state

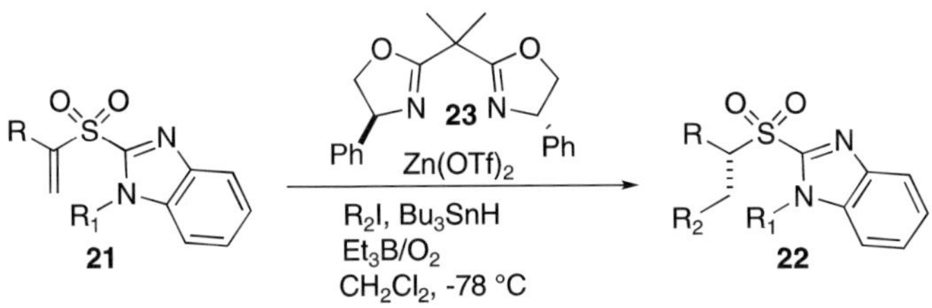

Entry	R	R_1	R_2	Yield (%)	ee (%)
1	Me	$C_6H_5CH_3$	*t*-Bu	88	56
2	Me	3,5-$Me_2C_6H_3CH_2$	*t*-Bu	97	82
3	Ph	3,5-$Me_2C_6H_3CH_2$	Et	75	68
4	Ph	Me	Et	99	91

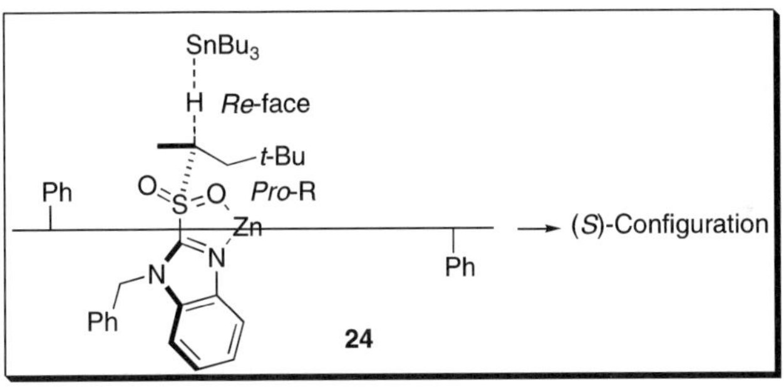

Scheme 7 Enantioselective hydrogen atom transfer to sulfones

involving tetrahedral zinc chelating the oxygen of the sulfonyl group and nitrogen atom of the benzimidazolyl group (**24**). Because of the steric repulsion between the neopentyl group and the phenyl group of the ligand, *pro*-R oxygen preferably coordinates with the zinc metal. A linear C – H – Sn geometry is preferred in the intermolecular hydrogen atom transfer from Bu_3SnH; thus, it approaches from a direction antiperiplanar to the bulky benzimidazolyl group (**24**). The opposite stereochemistry is merely a result of difference in the priority of groups and not because of the occupancy of different positions of the groups in the transition state.

A recent report by Sibi et al. demonstrated chiral Lewis acid mediated conjugate additions to dehydroalanines followed by enantioselective H-atom transfer to provide a variety of α-amino acid derivatives (Scheme 8) [25]. The chiral Lewis acid system derived from $Mg(ClO_4)_2$ and ligand **19** gave the best ee's. The intermediate obtained by the addition of a variety of nucleophilic radicals to **25** underwent H-atom transfer with good selectivity. It was shown that acetyl, α-alkoxyalkyl, primary alkyl, secondary alkyl, and cycloalkyl radical additions all give good selectivity in H-atom transfer (see entries 1–6). An exception to this trend was the reaction with bulky *tert*-butyl radical which gave low selectivity. This decrease in selectivity was attributed to the bulky

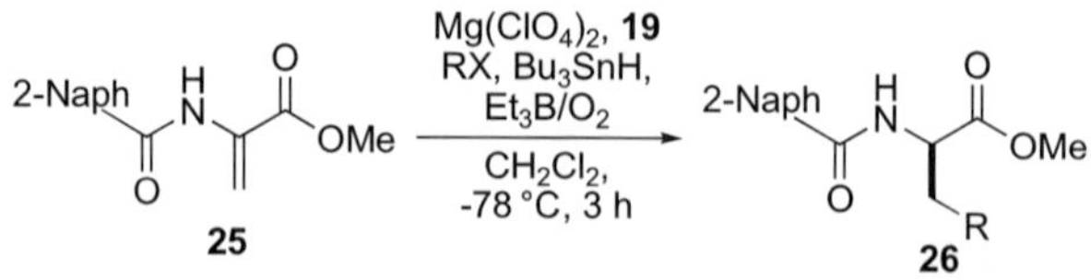

Entry	RX	Yield (%)	ee (%)
1	AcBr	76	80
2	$MeOCH_2Br$	71	65
3	EtI	72	85 (*R*)
4	*i*-BuI	76	79
5	*i*-PrI	62	83 (*R*)
6	*c*-HexI	62	55 (*R*)
7	*t*-BuI	54	27 (R)

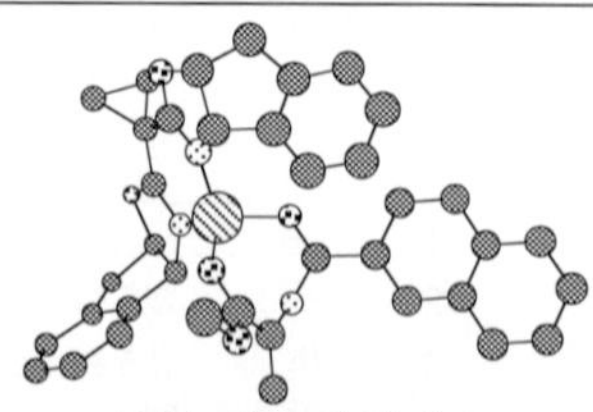
27 7-Membered Chelate *Re*-Face Shielding Observed

Scheme 8 α-Amino acids from dehydroalanines via enantioselective H-atom transfer

tert-butyl group and chiral Lewis acid shielding opposite faces resulting in reactions occurring from a mono-coordinated or non-complexed substrate. The enantioselective H-atom transfer follows the conjugate addition, and it is assumed that the structure of the intermediate radical resembles the starting complex. On the basis of these results a conjugate addition to a seven-membered chelate in the ternary complex (**27**) followed by a H-atom transfer was proposed. This is consistent with the observed stereochemistry.

Enantioselective synthesis of β-amino acids is important as they are present in various natural products and in many biologically active compounds [26, 27]. Several methods exist for the enantioselective synthesis of β-substituted β-amino acids (β^3-amino acids); however, synthesis of α-substituted β-amino acids (β^2-amino acids) is very limited [28, 29]. A report on highly enantioselective hydrogen atom transfer reactions to synthesize β^2-amino acids (Scheme 9) has recently been described [30].

Conjugate radical addition of various nucleophilic radicals followed by hydrogen atom transfer to β-amino acrylates in the absence of any Lewis acid proceeded to give high yields of the desired products. A Lewis acid/ligand screen found MgI_2 and bis-oxazoline ligand **19** to be the optimal catalyst system for the enantioselective reaction. Good yields and moderate selectivity were obtained when substrate **28** with methyl ester substituent was used (Scheme 9, entries 1 and 2). The catalytic reactions gave poor selectivity, which indicated that the background reaction competes with the catalyzed reaction. Interestingly, with a change of the ester substituent to a bulky *t*-butyl group, good selectivity was obtained (entries 3–5). The selectivity was found to be directly proportional to the size of the radicals added, and bulky radicals provided > 90% enantioselectivity (entry 5). The selectivity was remarkable given that the rate of background reaction was very fast. Interestingly, using substoichiometric amounts of chiral Lewis acid (30 mol %), higher selectiv-

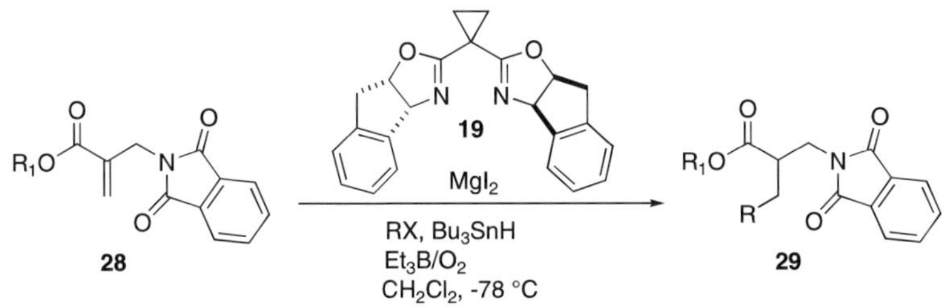

			100 mol% CLA		30 mol% CLA	
Entry	R_1	R	Yield (%)	ee (%)	Yield (%)	ee (%)
1	Me	*i*-Pr	91	40	95	20
2	Me	*t*-Bu	81	20	-	-
3	*t*-Bu	Et	82	36	83	62
4	*t*-Bu	*i*-Pr	91	62	95	84
5	*t*-Bu	*t*-Bu	85	92	88	71

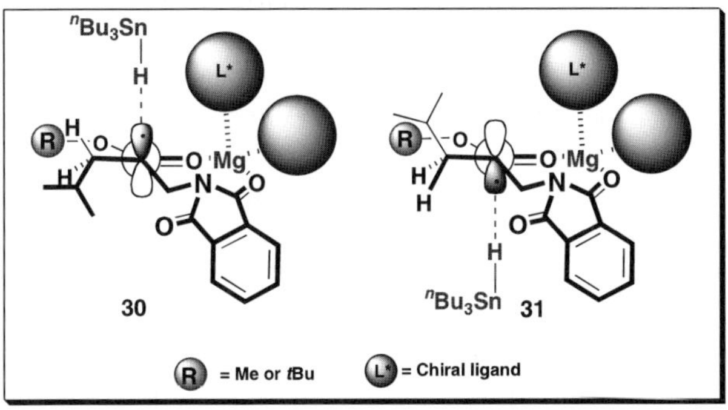

Scheme 9 Synthesis of β^2-amino acids by enantioselecive hydrogen atom Transfer

ity was obtained compared with 100 mol% of the catalyst when ethyl and isopropyl radicals were added (entries 3 and 4).

The absolute stereochemistry of the products from isopropyl radical addition was found to be (*S*) for both the ester substrates. An eight-membered chelate model was proposed to explain the absolute configuration of the products and some of the above observations (Scheme 9) [30]. The conformation of the ester substituent is dependent on its size and is controlled by the ligand. The bulky *t*-butyl ester is predominantly in the *s-trans* orientation. The conformation of the methyl ester is not fixed but still exists predominantly in the *s-trans* arrangement. After the addition of radical from the top face, the face selectivity of the hydrogen atom transfer is dependent upon the size of the ester substituent and that of the radical fragment. Steric interactions between the *t*-butyl ester and the radical fragment in the complex force the bond to rotate and adopt the orientation as shown in **30**. Hydrogen atom transfer then occurs *anti* to the radical fragment. The steric interactions between the methyl group of the ester and the radical fragment are less demanding;

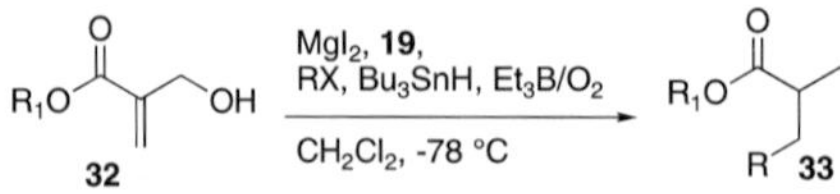

Entry	R_1	R	Yield (%)	ee (%)
1	Me	*i*-Pr	69	88
2	Me	*c*-Pent	82	90
3	Me	*t*-Bu	77	92
4	*t*-Bu	*i*-Pr	53	55
5	*t*-Bu	*c*-Pent	62	69
6	*t*-Bu	*t*-Bu	80	53

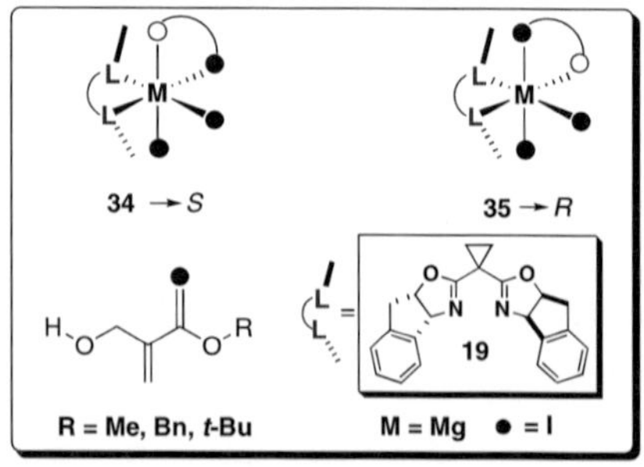

Scheme 10 Synthesis of formaldehyde aldol products

hence, low selectivity is obtained although reactions occur predominantly through **30**.

A novel synthetic pathway for the synthesis α-substituted aldol products from simple α-hydroxymethyl acrylates has been developed by Sibi and co-workers [31–38]. Using low temperature radical conditions, substrates **32** react in their free alcohol form without the need for protection, providing access to enantioenriched formaldehyde aldol products (**33**) [39]. Scheme 10 shows the radical addition-hydrogen atom transfer to the substrates under stoichiometric chiral Lewis acid conditions. It was observed that methyl ester substituents consistently gave good selectivity and bulky radicals provided very high enantioselectivity (entries 1–3), while *t*-butyl esters provided only moderate selectivity (entries 4–6). Catalytic conditions were not very efficient, and the low selectivity was attributed to the high rate of background reaction.

Surprisingly, the absolute configuration was found to be complementary for both the substrates. The authors speculate that both the substrates react via octahedral magnesium complexes, but coordination in either of two orientations is possible depending on the size of the ester substituent [39]. With the methyl ester, the ester carbonyl coordinates *trans* to one of the ligand nitrogen atoms (**34**), as shown in Scheme 10. The orientation of the bulky *t*-butyl substituent reverses and then reacts via complex **35** as shown in which the hydroxy group rather than the ester carbonyl occupies the position *trans* to the ligand nitrogen. The *t*-butyl substrate reacts primarily via complex **35**; however, stereochemical erosion results from significant competing reaction via complex **34**.

2.1.2 Chiral Hydrogen Bonding Agent

An interesting intramolecular radical cyclization followed by enantioselective hydrogen atom transfer has recently been reported (Scheme 11) [40]. This reaction is carried out in the presence of a chiral complexing agent **38**, which

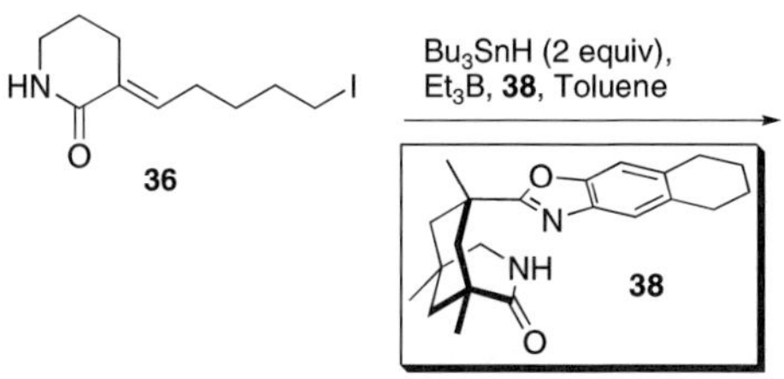

Entry	T (°C)	Equiv (38)	Et_3B (mol%)	Yield (%)	ee (%)
1	25	-	50	83	-
2	-10	2.5	10	79	55
3	-78	1.0	20	91	40
4	-78	2.5	50	84	41
5	-78	2.5	20	81	84

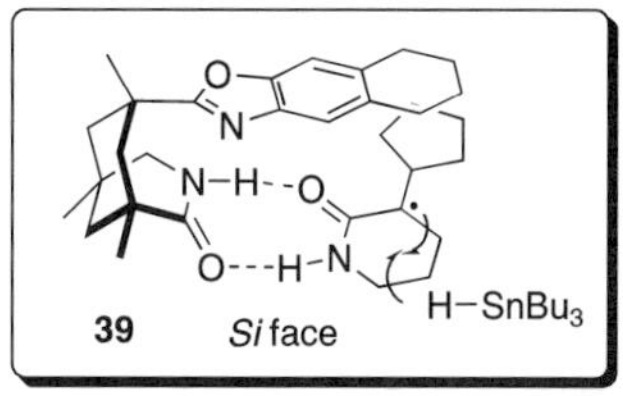

Scheme 11 Enantioselective H-atom transfer reaction with hydrogen bonding catalyst

can hydrogen bond to an appropriate acceptor. The authors studied the enantioselectivity of the reductive cyclization of 3-(ω-iodoalkylidene)piperidin-2-ones (**36**). The observed selectivity was determined to be dependent on three factors. First, the temperature needed to be very low (– 78 °C). Secondly, lower amounts of radical initiator (Et_3B) gave improved enantioselectivities (entries 4 and 5). Another interesting reaction parameter was the need for a large excess of the chiral source (**38**) (entries 3 and 5). A proposed model for the observed stereochemistry is shown in Scheme 11. The *re* face is shielded by the tetrahydronaphthalene moiety and the enantioselective hydrogen atom transfer proceeds from the more accessible *si* face of the prochiral radical **39**.

2.1.3 Memory of Chirality

Rychnovsky et al. considered the formation of achiral conformers from chiral molecules and trapping the prochiral radical with a hydrogen atom donor based on memory of chirality (Scheme 12) [41]. The photo-decarboxylation of optically active tetrahydropyran **40** leads to an intermediate **43**, which now does not contain a stereocenter. If the intermediate **43** can be trapped by some hydrogen atom source before ring inversion takes place, then an optically active product **41** will be formed. This is an example of conformational memory effect in a radical reaction. It was reported that the radical inversion barrier is low (≤ 0.5 kcal/mol) while the energy for chair flip **43** ⇎ **44** is higher (5 to 10 kcal/mol).

The photochemical reduction of Barton ester **40** is depicted in Scheme 12. A series of hydrogen atom donors were screened. A stoichiometric amount of benzenethiol at – 78 °C provided the product in 86% ee (entry 3). This implies that, in the presence of an efficient hydrogen atom donor, radical trapping is competitive with the ring/radical inversion, generating an enantiomeri-

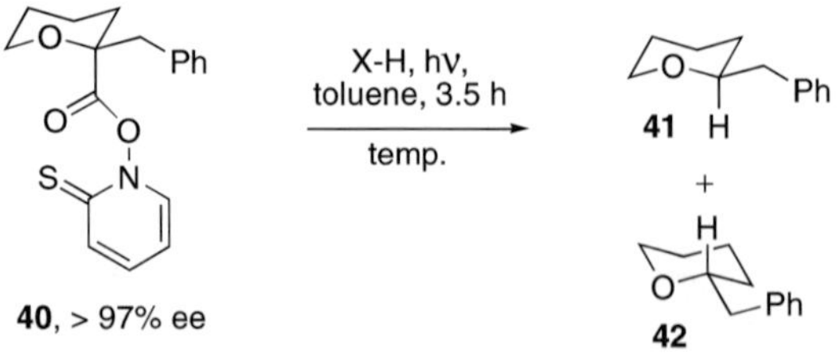

Entry	Donor	[X-H] (M)	Temp (°C)	Yield (%)	ee (%)
1	Bu_3SnH	1.0	-78	70	3
2	*t*-BuSH	1.0	-78	83	26
3	PhSH	1.0	-78	92	86
4	PhSH	0.5	-78	75	70
5	PhSH	0.5	-40	50	40
6	PhSH	0.5	0	77	15
7	PhSH	0.5	22	72	8
8	PhSeH	0.05	-78	28	35

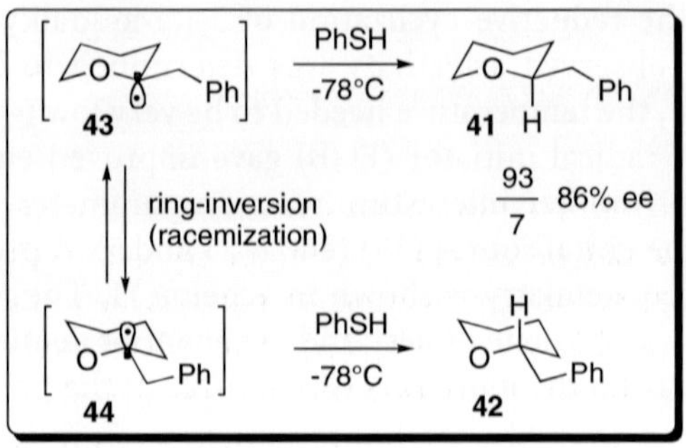

Scheme 12 Memory of chirality: reduction of Barton esters

cally enriched compound **41**. Substoichiometric amounts of benzenethiol at – 78 °C furnished the reduced product with low selectivity (entry 4). This observation can be explained by the fact that reduced thiol concentration would allow more time for the racemization. As observed with most of the asymmetric reactions, the selectivity decreases upon increasing the temperature (entries 5–7). Tributyltin hydride and benzeneselenol proved to be poor hydrogen atom donors for obtaining high selectivities in this reaction (entries 1 and 8).

Another report by Rychnovsky et al. explored the potential of chirality transfer in the transannular cyclization of cyclodecene **45** [42]. They proposed a radical deoxygenation of **45**, which produces an intermediate cyclodecenyl radical that can cyclize in a 5-*exo* fashion to yield 5,7-fused bicycle **48** (Scheme 13). The potential for the optically enriched radical precursor **45** to undergo enantioselective cyclization is dependent on the rate of transannular cyclization. That is, if the radical generated from optically pure

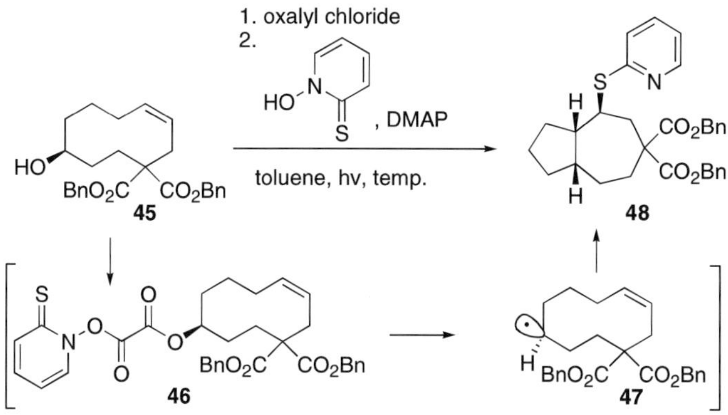

Entry	Temp. (°C)	Yield (%)	ee (%)
1	23	88	26
2	0	67	58
3	-15	51	68
4	-35	43	68

Scheme 13 Memory of chirality in radical cyclizations

45 underwent cyclization before conformational interconversion, enantioenriched products would be obtained. The deoxygentation/radical cyclization procedure was carried out by treating **45** with oxyalyl chloride followed by *N*-hydroxy pyridine thione and 4-DMAP. The mixed oxalate **46** was formed in situ and photolysis of the reaction mixture produced bicyclo[5.3.0]decane **48**. Low selectivities were observed at room temperature (entry 1), but moderate enantioselectivities could be achieved at lower temperatures (entries 3 and 4). One explanation for the erosion of selectivity was the presence of two conformational isomers in the starting substrate.

2.1.4 Chiral Reagent

Hydrogen atom transfer reactions can also proceed via the use of a chiral reagent (Scheme 14). This process will be successful only if the reagent can differentiate between the enantiotopic faces of a radical in diastereomeric transition states. The geometry of the approach in a hydrogen atom transfer reaction is linear, and hence, sterically differentiating groups adjacent to the hydrogen transferred should induce stereocontrol. However, the elements should not be too bulky to hinder the approach of the reagent to the prochiral radical.

More than two decades ago, Ohno and co-workers synthesized optically active nicotinamide **55**, which was considered a chiral model of NAD(P)H [43]. The model compound afforded high enantiospecificity in

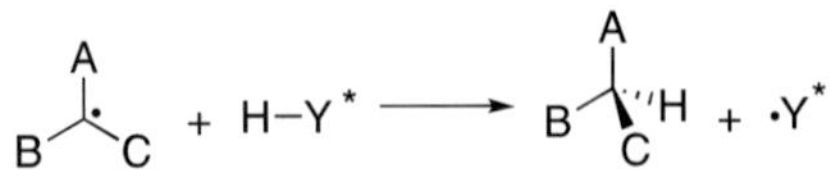

49 50 51 52

Scheme 14 Hydrogen atom transfer through a chiral tin hydride

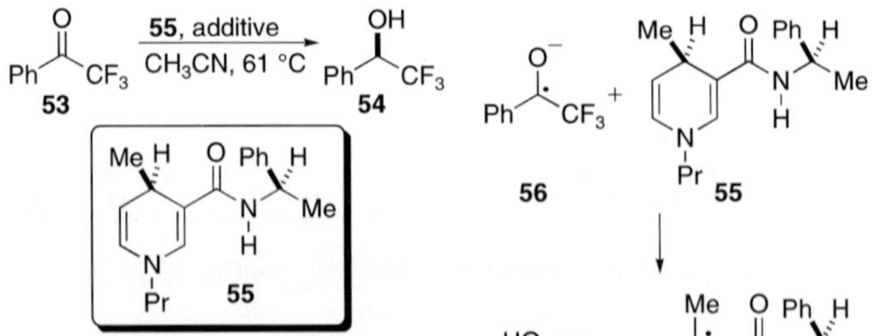

Entry	Additive (%)	Yield (%)	ee (%)
1	-	43	66
2	DNB (4)	13	66
3	AIBN (3)	82	68

Scheme 15 Reduction of ketones: chiral nicotinamides

the reduction of certain carbonyl compounds and provided ~ 70% ee. Later, Tanner and Kharrat studied this reaction in detail and found that the reduction of ketone **53** was initiated in the presence of 2,2′-azabisisobutyronitrile (AIBN), and was inhibited in the presence of an efficient electron acceptor, *m*-dinitrobenzene (DNB) (Scheme 15, entry 2) [44]. The non-initiated, partly inhibited, and the initiated reactions gave the same optical purity of alcohol **54**. However, the different yields of **54** imply that these reactions proceeded through a radical pathway. The mechanism involved the transfer of a hydrogen atom from the chiral nicotinamide to ketyl radical **56** as shown in Scheme 15. Although a stereocenter is lost, the new radical still maintains the other stereocenter; dihydropyridyl radical **57** then propagates the chain.

2.1.5 Chiral Stannanes and Germanes

There can be two kinds of chiral tin reagents: tin chiral and C-chiral. Early reports of chiral tin hydride involved transfer of chirality via a chiral tin center [45–47]. These tin hydrides were prone to racemization. Thus, chiral carbon-based ligands attached to the tin center were synthesized to minimize racemization. The first chiral tin hydride containing a C_2-symmetric binaphthyl substituent was reported by Nanni and Curran (Scheme 16) [48]. α-Bromoketone **58** was reduced by chiral tin hydride **59** (R_3 = Me), where the reactivity and selectivity was dependent on the reaction conditions (entry 4).

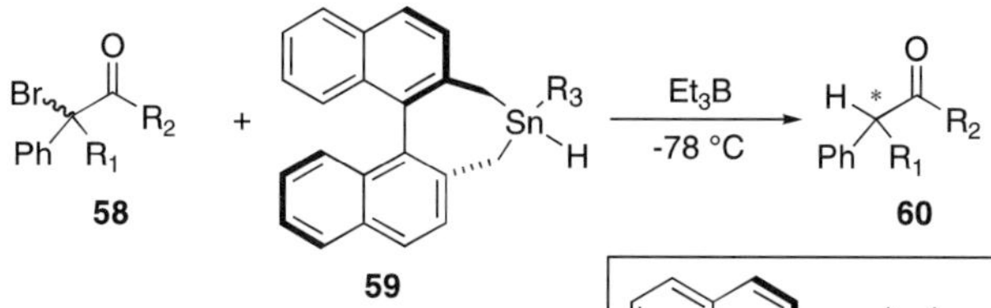

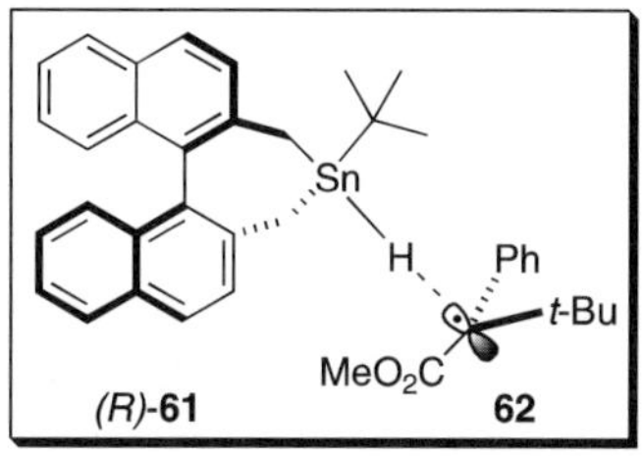

Entry	R_1	R_2	R_3	ee (%)
1	*t*-Bu	OMe	*t*-Bu	52
2	*i*-Pr	OMe	*t*-Bu	34
3	Me	OMe	*t*-Bu	04
4	Me	Ph	Me	41

Scheme 16 Reduction of α-bromocarbonyls: chiral tinhydride

The selectivity was much better when excess triethylborane was used as an initiator, although the yields were modest. Employing AIBN as an initiator, the reactivity improved, but selectivity decreased.

Around the same time, Metzger and co-workers synthesized *t*-butyl substituted binaphthyl tin hydride **59** independently using an alternate procedure and employed them in the reduction of bromoester **58** (Scheme 16) [49]. The reaction was highly efficient providing up to 52% enantioselectivity (entry 1). A full account of this work has been recently published [50].

Scheme 16 summarizes the results obtained by enantioselective radical reduction of α-bromoester by chiral binaphthyl-derived tin hydride. The reactions were generally performed at – 78 °C. An increase in the temperature resulted in the lowering of the selectivity. All reactions mediated by (*S*)-configured chiral tin hydride showed an (*R*)-selective preference in the product. The use of the opposite enantiomer of the chiral stannane resulted in a quantitative reversal of the selectivity (not shown). The selectivity remained modest on addition of magnesium Lewis acids. These reductions were also feasible when a catalytic amount of chiral tin hydride (1 mol %) was employed in combination with an excess of achiral hydride $NaCNBH_3$, providing similar results.

Metzger and co-workers have also described a reduction of α-bromoesters by chiral tin hydrides containing a diastereomeric mixture of 2-[(1-dimethylaminoalkyl)phenyl] (DAAP) ligands [51]. The observed enantioselectivities were dependent on the tin hydride used and on the substituents attached to the radical center.

Thomas and co-workers have examined camphor-derived chiral stannanes recently [52]. However, poor selectivity (< 5% ee) was obtained for the reduction of bromoketones in the absence of any Lewis acid.

Schiesser and co-workers have evaluated the use of achiral and chiral Lewis acids in chiral hydrogen atom donor mediated reactions [53–59]. The Lewis acid additives greatly enhance the enantioselectivity in free radical reductions of halo esters and ketones in the presence of menthol or cholic acid-derived chiral stannanes (Scheme 17). Among the Lewis acids used, the achiral Lewis acids, namely BF_3 and Cp_2ZrCl_2 (data not shown), provided moderate selectivity (entry 2) with substrate **63**. In the presence of Jacobsen's catalyst **65**, much higher selectivity was obtained (entry 3). The enantiomeric catalysts **65** and **66** gave similar selectivity with the same absolute configuration of the product (entries 3 and 4), implying that the Lewis acid only provides the steric bulk at the coordinating carbonyl and does not have an influence on the control of the face selection in the hydrogen atom-transfer process.

Although the Lewis acid greatly enhances the selectivity, the transfer of chirality is derived from the chiral ligand on the stannane. These deductions are supported by the fact that when stannane **67** is used, the ee of the product increases from 4% in the absence of a Lewis acid to 46% in the presence of achiral Lewis acid (Cp_2ZrCl_2) for substrate **63**. When the enantiomer of **67** was used as the reductant, the product was obtained with the opposite configuration, which also supports the above-mentioned presumption.

The above methodology should enable control of product configuration by mere change in the reagent. Preparation of natural and unnatural amino acid

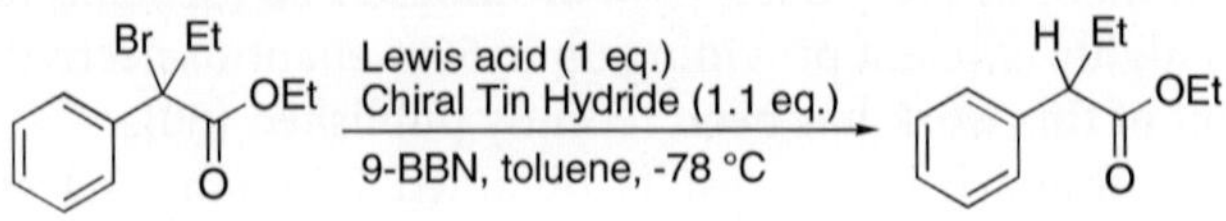

63 **64**

Entry	Lewis Acid	Stannane	Yield[a] (%)	ee (%)
1	None	**67**	81	41(S)
2	BF_3	**67**	68	20 (S)
3	**65**	**67**	75 (71)	86 (S)
4	**66**	**67**	69	84 (S)
5	**65**	**68**	68	72 (S)
6	**65**	**68**	N/A	>99 (S)

[a] GC conversion; numbers in paranthesis are isolated yields.

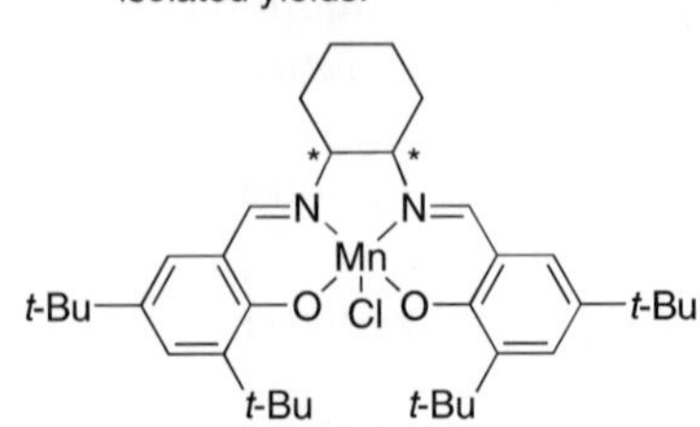

65 *S,S*
66 *R,R*

men$_2$PhSnH
67

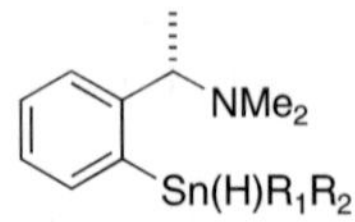

68 R_1=R_2=men

Scheme 17 Lewis acids as additives in reductions

derivatives was next attempted (Scheme 18). A large enhancement in selectivity was observed when magnesium Lewis acids were used as compared to reactions in the absence of a Lewis acid; however, substrate **69** gave poor selectivity even in the presence of magnesium Lewis acids (entries 1 and 2). The poor selectivity was attributed to the steric similarity of the methyl and ethyl groups attached to the prochiral radical; of particular interest was the addition of zeolites such as zinc, calcium, and magnesium silicates, which also resulted in excellent enantioselectivities (data not shown).

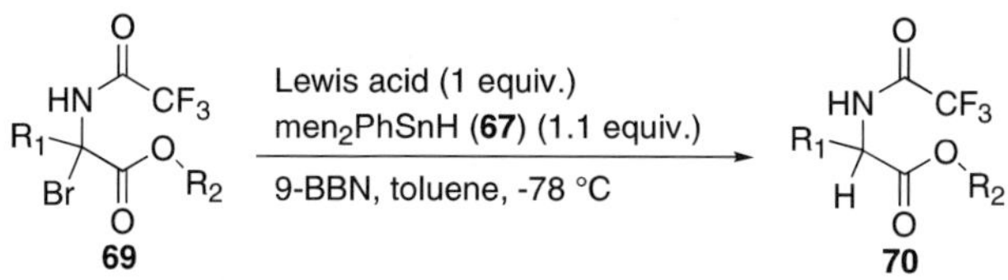

Entry	R1	R2	Lewis Acid	Yield (%)	ee (%)
1	Et	Me	None	n.d.	00
2	Et	Me	$MgBr_2$	n.d.	10
3	Ph	Me	None	70	05
4	Ph	Me	$MgBr_2$	78	99 (*R*)
5	Ph	Me	$ZnCl_2$	n.d.	14 (*S*)
6	*t*-Bu	Bn	$MgBr_2$	65	99 (*S*)
7	*t*-Bu	Bn	$Mg(OAc)_2$	95	99 (*S*)
8	*t*-Bu	Bn	MgI_2	96	99 (*S*)

Scheme 18 Synthesis of α-amino acid derivatives: use of chiral stannane

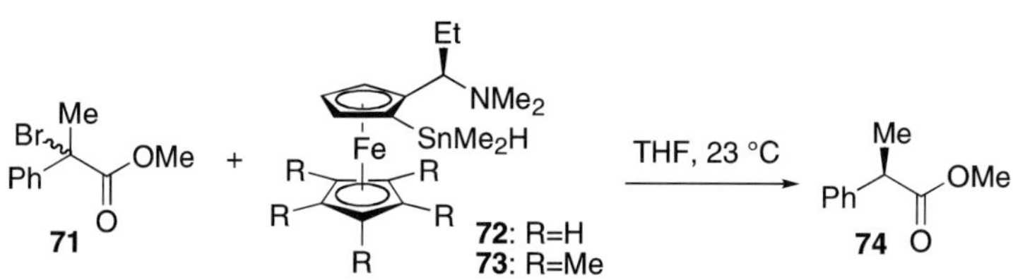

Entry	Ligand	Time	ee (%)	Conversion
1	**73**: (1.0 equiv)	20 h	87%	65%
2	**72**: (1.5 equiv)	5 h	9%	100%

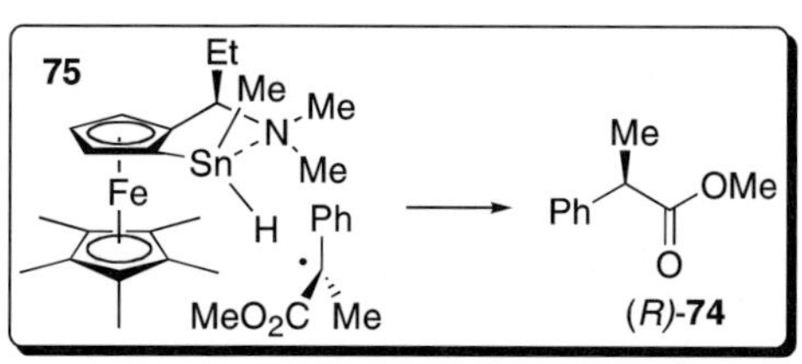

Scheme 19 Ferrocene-derived chiral stannane

Recently, Kang and Kim developed new chiral ferrocenyl tin hydride derivatives **72** and **73** (Scheme 19) [60]. The authors screened the new chiral reagent in the reduction of α-bromoesters. Using one equivalent of **73** good ee's were obtained for ester **71**. One drawback for this reagent, however, is the lengthy synthetic route for its preparation.

Researchers have studied alternative chiral metal hydride reagents to carry out enantioselective free radical transformations. Although germanium hydrides are slower hydrogen donors than the tin hydride analogs, they still have favorable rate constants as radical chain propagating agents. Several groups have studied these chiral germanium hydride species with moderate to good success in achieving high enantiomeric excess [61–63].

2.1.6 Polarity-Reversal Catalysis

A majority of radical addition occurs with electron-poor alkenes using alkyl halides in the presence of Bu_3SnH. These reactions are feasible due to a proper matching between the radical acceptor and the donor. However, when the alkene is electron-rich and since simple alkyl radicals are considered as nucleophilic, the reaction is not a practical method for carbon–carbon bond formation. By applying the concept of polarity-reversal catalysis, an additional reagent is introduced which alleviates the mismatch between the partners and makes the reaction feasible. A few examples illustrating this concept have been described in this review.

The kinetic resolution of racemic *trans* ester **76a** using catalytic amounts of chiral amine-borane **78** and di-*t*-butyl peroxide as initiator under photolytic conditions at –74 °C provided the enantioenriched (*R*,*R*) product in 74% ee after 52% consumption of the racemate [64–67]. For the ester **76b**, (*R*,*R*) product in 97% ee was isolated after 75% consumption at –90 °C (Scheme 20).

Abstraction of α-hydrogen atom from **77** to generate another electrophilic radical is not possible in the absence of **78**. The reaction of electrophilic *t*-butoxy radical with **78** generates a highly nucleophilic radical **79**, which is then forced to abstract the α-hydrogen atom from **76**. The amine-borane catalyst acts as a hydridic polarity reversal catalyst. Enantioselective hydrogen atom abstraction by chiral amine-boryl radical from **76** gives rise to radical **77**, which then decomposes, thereby enhancing the selectivity of the residual ester. The selectivity for this reaction can be explained by the transition state, **80**, for the hydrogen atom abstraction step.

Roberts and co-workers have employed a number of chiral carbohydrate-derived thiols as polarity reversal catalysts in the radical hydrosilylation of electron-rich prochiral alkenes [68–70]. In these thiols, the SH group is attached to the anomeric carbon atom. Scheme 21 demonstrates the non-catalyzed reaction and in step **b**, the hydrogen atom transfer from the silane

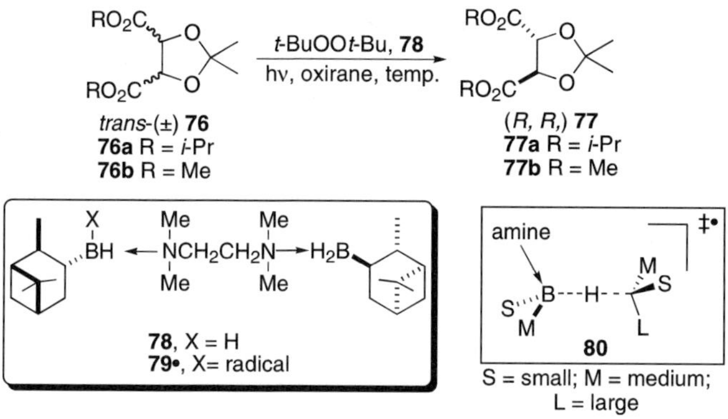

76	Temp. (°C)	Conversion (%)	ee (%)
76a	-74	52	74
76b	-90	75	97

$$t\text{-BuO}\bullet + \mathbf{78} \xrightarrow{\text{fast}} t\text{-BuOH} + \mathbf{79}\bullet \quad (1)$$

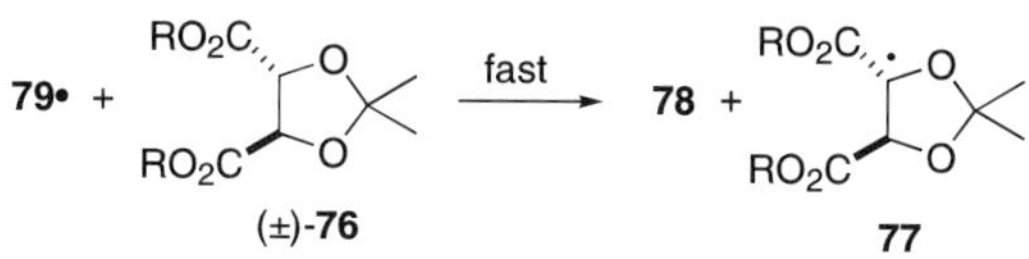
(2)

Scheme 20 Chiral amine-borane catalyzed kinetic resolution

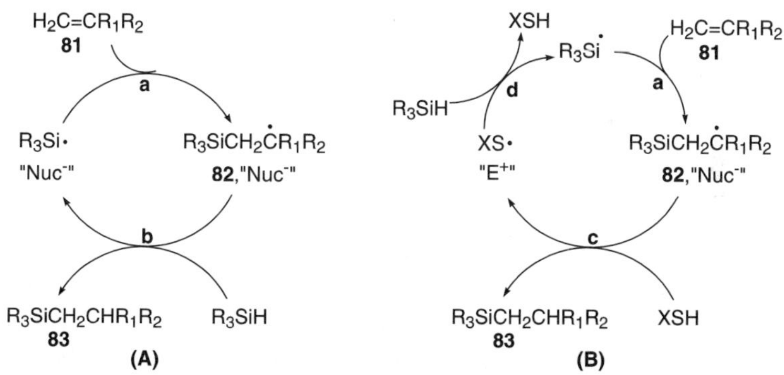

Scheme 21 Hydrosilylation: thiols as polarity reversal catalysts

to the carbon-based radical is slow as both, the hydrogen atom donor and the acceptor, are nucleophilic. In the thiol-catalyzed cycle (**B**), the slow step is replaced by faster propagation steps **c** and **d**. The SH group of thiol will provide electron deficient hydrogen, which favors hydrogen atom transfer to the nucleophilic alkyl radicals that are formed by addition to electron-rich alkenes. In principle, if **82** is a prochiral radical and the thiol is optically active, then step **c** should provide an enantioenriched product.

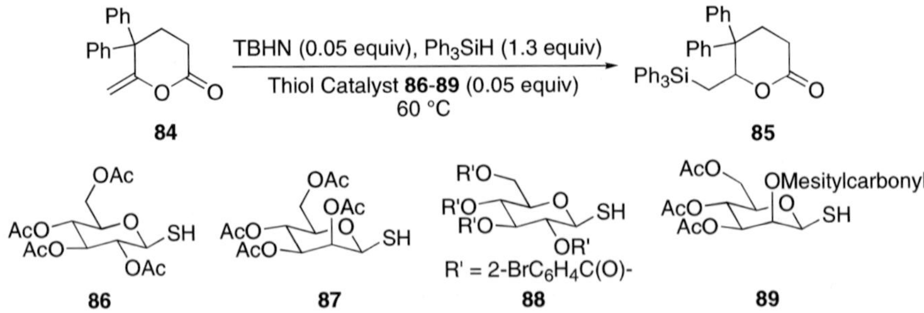

Entry	Thiol	Solvent	Yield (%)	ee (%)
1	**86**	Dioxane	88	80
2	**86**	Benzene	92	86
3	**86**	H + D (5:1)	93	87
4	**87**	H + D (5:1)	90	95
5	**88**	H + D (4:1)	73	85
6	**89**	H + D (4:1)	87	94

H = Hexane; D = Dioxane

Scheme 22 Hydrosilylation of lactones

Optically active organosilanes (**85**) were synthesized from methylene-δ-lactones **84** via this method (Scheme 22). The sugar-derived thiols (5 mol %) were used in the presence of a slight excess of silane and with di-*tert*-butyl hyponitrite (TBHN, 5 mol %) as an initiator. For substrate **84**, under a variety of conditions using different silanes, a moderate level of selectivity was achieved (data not shown). The diphenyl-substituted precursor, **84**, provided much better yields and selectivity for the addition of triphenylsilane (Ph_3SiH) followed by hydrogen atom transfer. It was found that a mixture of hexane/dioxane (H/D, 5 : 1 or 4 : 1) with thiol catalysts **86–89** furnished highest selectivities with substrate **84** (entries 3 and 6). These optically active adducts formed by hydrosilylation can be oxidatively desilylated to provide useful organic products. Other systems studied using chiral thiol catalysts include kinetic resolution of silanethiyl radicals and reductive carboxyalkylation of electron-rich alkenes; however, modest enantioselectivity was obtained in these reactions [71, 72].

2.2 Halogen Atom Transfer

Halogen atom transfer reactions involve homolysis of a C – X or a X – X bond in a neutral molecule and transfer of both radical components to unsaturated functional groups. There is atom economy in such processes and they provide functionality for further transformations [73].

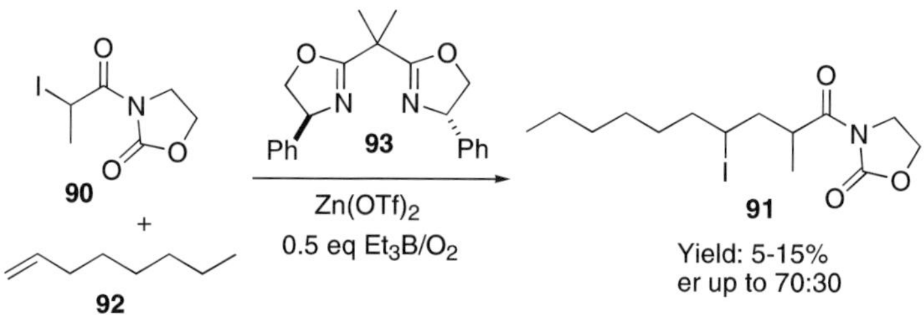

Scheme 23 Atom transfer addition of α-iodoimide **90**

Ruthenium complexes are capable of catalyzing halogen atom transfer reactions to olefins. This has been illustrated in the enantioselective atom transfer reactions of alkane and arene-sulfonyl chlorides and bromotrichloromethanes to olefins using chiral ruthenium complexes. Moderate ee's up to 40% can be achieved for these transformations [74–77]. These specific reactions are believed to follow a radical redox transfer chain process.

Chiral Lewis acid promoted atom transfer reaction (Kharasch reaction) of α-halo oxazolidinone imide **90** and 1-octene **92** has been reported by Porter et al. (Scheme 23) [78]. The enantioselective atom transfer utilizing $Zn(OTf)_2$ and phenyl bisoxazoline ligand **93** as a chiral Lewis acid. The yields of the products, however, were quite low ranging from 5–15% and only moderate enantioselectivities were achieved (up to 40%).

Arylation of activated double bonds with diazonium salts in the presence of copper catalysts is known as the Meerwin reaction. The reaction is postulated to either proceed through an organocopper intermediate or through a chlorine atom transfer from chiral CuCl complex to the α-acyl radical intermediate. Brunner and Doyle carried out the addition of mesityldiazonium tetrafluoroborate with methyl acrylate using catalytic amounts of a Cu(I)-bisoxazoline ligand complex and were able to obtain 19.5% ee for the product (data not shown) [79]. Since the mechanism of the Meerwin reaction is unclear, it is difficult to rationalize the low ee's obtained and to plan for further modifications.

2.2.1 Cyclization

Highly enantioselective atom transfer radical cyclization reactions catalyzed by chiral Lewis acids have been reported by Yang et al. [80]. Two main advantages of these enantioselective cyclizations include installing multiple chiral centers and retaining a halogen atom in the product, which allows for further functionalization.

Scheme 24 shows the atom transfer radical cyclizations of unsaturated β-keto esters **94** using $Mg(ClO_4)_2$ and chiral ligand **96**. It was found that toluene as a solvent generally gave higher enantioselectivities than CH_2Cl_2

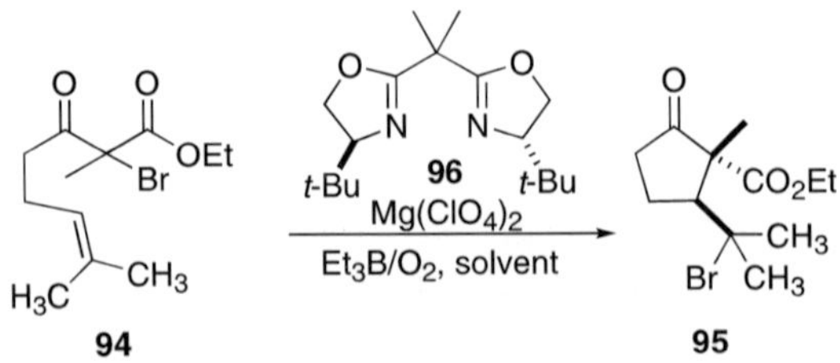

Entry	Catalyst (equiv.)	Solvent	Time (h)	Yield (%)	ee (%)
1	1.1	CH_2Cl_2	7.5	68	71
2	1.1	toluene	5	67	94
3[a]	0.5	toluene	7	65	93
4[b]	1.1	toluene	9	53	21

[a] MS 4Å (500mg/mmol substrate)
[b] 1.0 equiv. of water was added

Scheme 24 Atom transfer cyclization

(see entries 1 and 2). Both 5-exo and 6-exo (not shown) cyclization proceeded uneventfully. One notable observation was the addition of activated 4 Å molecular sieves, which proved to enhance ee's and allow for the use of substoichiometric amounts of chiral Lewis acid (entry 3). The molecular sieves are thought to act as a drying agent: the addition of 1.0 equivalent of water drastically reduces the selectivity and cyclization rate of **94** (see entries 2 and 4). Catalytic loading of the chiral Lewis acid showed nearly identical efficiency as stoichiometric amounts of chiral Lewis acid with respect to both chemical yields and enantioselectivities (compare entries 2 and 3).

The high selectivity can be explained by the model shown in Scheme 25. Because of the steric bulk of the *t*-butyl groups of bisoxazoline ligand **96**, *re*-face cyclization (transition states **97** and **98**) should be favored over *si*-face

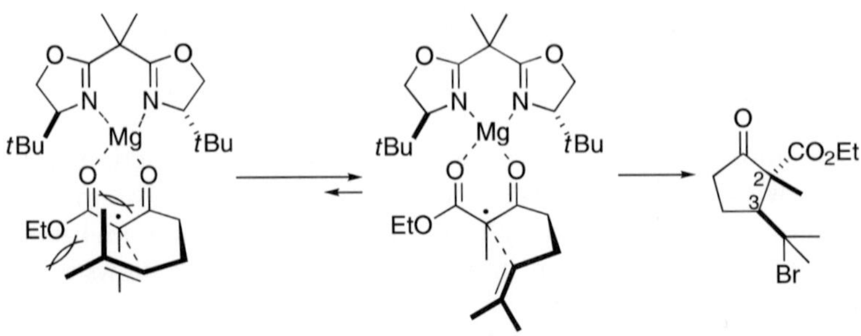

97
disfavored

98
favored

(2*R*, 3*S*)-**95**

Scheme 25 Model for selectivity in 5-*exo* cyclization

cyclization (structures not shown). Transition state **98** results in the lowest overall steric interaction and leads to product **95** with (2*R*, 3*S*) configuration where the ester group on C2 and the alkyl group on C3 are *trans* to one another.

Enantioselective tandem radical cyclization reactions can provide access to highly functionalized polycyclic compounds with multiple stereocenters. Yang et al. have recently reported the first Lewis acid catalyzed enantioselective atom transfer tandem cyclization reaction (Scheme 26). It was found that the enantioselective tandem cyclization of **99** using $Mg(ClO_4)_2$ and chiral ligand **96** in CH_2Cl_2 gave poor ee's (entry 1). Molecular sieves slightly increased the ee but reduced the chemical yield by half (entry 2). Substrate **100** in toluene at higher temperatures gave good enantioselectivities but still poor yields were obtained (entries 3 and 4).

Cyclization of substrate **99** could also be performed with $Yb(OTf)_3$ as the Lewis acid in the presence of several chiral ligands (Scheme 27) [81]. The best results were obtained using the **105**/Yb complex in CH_2Cl_2, which gave 60% yield of **104** with 66% ee (entry 1). It is interesting to note that the addition of 4 Å molecular sieves gave nearly complete reversal of enantiofacial selectivity in the tandem radical cyclization along with an increase in the reduced product (compare entries 1 and 2). Toluene as the solvent gave lower yields of **104** and ee's compared to methylene chloride.

Scheme 28 explains the stereochemical outcome from the tandem radical cyclization in the presence of the [Yb(Ph-pybox)$(OTf)_3$] (pybox = 2,6-bis(2-oxazolin-2-yl)pyridine). The ytterbium complex **107** is shown in an octahedral geometry (with one triflate still bound to the metal) where *re*-face cyclization is favored due to the steric interactions of the substrate and the ligand's phenyl groups. The 6-*endo* cyclization takes place via a chair-like transition state to yield a tertiary radical **108** followed by a ring flip and

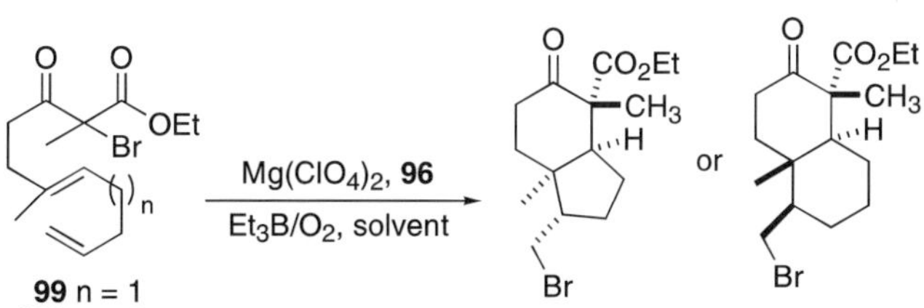

99 n = 1
100 n = 2

101 **102**

Entry	Substrate	T (°C)	Solvent	Product	Yield (%)	ee (%)
1	**99**	-78	CH_2Cl_2	**101**	41	13
2[a]	**99**	-78	CH_2Cl_2	**101**	24	33
3	**100**	-40	toluene	**102**	23	82
4	**100**	-20	toluene	**102**	16	84

[a] MS 4Å was added

Scheme 26 Tandem cyclizations using atom transfer additions

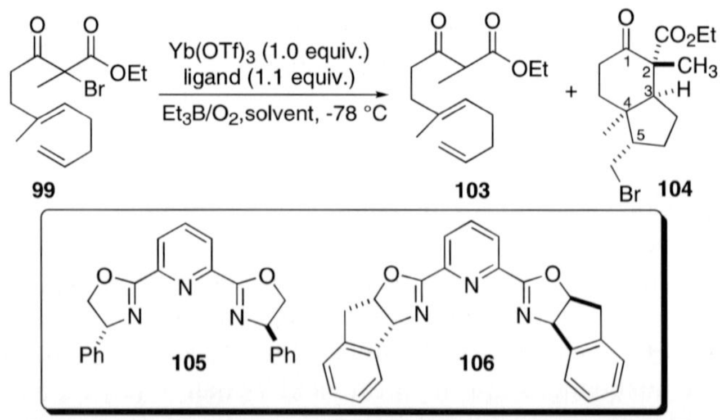

Entry	Ligand	Solvent	t (h)	Yield (%) [**103**]	Yield (%) [**104**]	ee (%)
1	**105**	CH_2Cl_2	15	23	60	66
2[a]	**105**	CH_2Cl_2	13	68	11	-56
3	**106**	toluene	10	64	17	-43

[a] MS 4Å was used

Scheme 27 Ytterbium mediated tandem cyclization

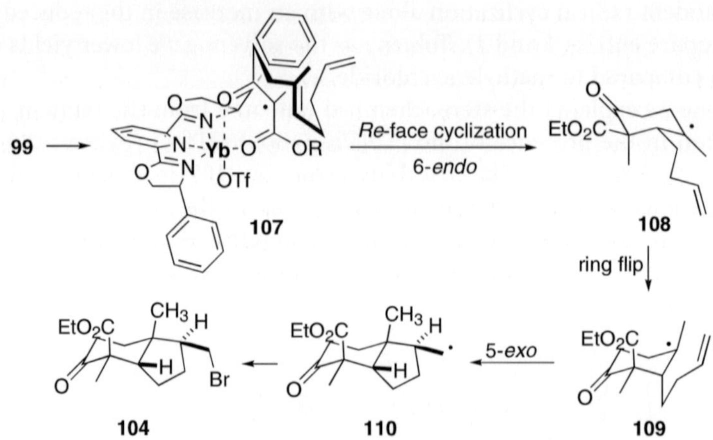

Scheme 28 Model/mechanism for $Yb(OTf)_3$ mediated cyclization

a 5-*exo* cyclization (**108** to **109** to **110**). The primary radical in **110** then abstracts a bromine atom from **99** to yield (2*R*, 3*S*, 4*S*, 5*S*)-**104**.

3 Reductive Alkylations

Addition of radicals to carbon–carbon or carbon–heteroatom multiple bonds followed by the trapping of resulting radicals with a hydrogen atom source leads to reduced products. A very favorable situation for catalytic processes exists here if the chiral Lewis acid modulates the reactivity of substrate suitably.

3.1 Additions to Imines

Glyoxylate imines have proven to be good substrates for the enantioselective ene and hetero Diels–Alder reactions. Radical addition to glyoxylate imines has been carried out with chiral Lewis acids. These reactions can provide optically active aliphatic α-amino acids. The radical methodology is advantageous since anionic nucleophiles do not distinguish the imine and the carboxylic esters and regioselectivity is not attained. The only other case where such selectivity is obtained is in the addition of allyl metal reagents. Naito et al. have utilized catalysts derived from **23** with various metal salts in addition of *iso*-propyl radical to **111** with limited success (Scheme 29) [82]. Among the Lewis acids evaluated, only magnesium bromide gave reasonable ee's. A tetrahedral model **113** has been proposed for the observed selectivity. Similar studies were carried out by Jørgensen et al. using chiral Lewis acid derived from Cu(I) and Tol-BINAP **116** with less success (Scheme 30) [83]. The use of Et_3B as the initiator at room temperature in the absence of tributyltin hydride seems novel in this study. At low temperature where the concentration of *i*-Pr radical is low, higher amounts of ethyl addition product (Et• obtained from Et_3B) was observed. At room temperature, the atom transfer step (Et• + *i*-PrI → *i*-Pr• + EtI) to produce the *i*-Pr radical is more efficient and the use of tributyltin hydride can be avoided. This also restricts this particular process to alkyl iodides; other halo alkanes have much lower halogen transfer rates.

Entry	Lewis acid	Yield (%)	ee (%)
1	$Zn(OTf)_2$	73	10 (*R*)
2	$Yb(OTf)_3$	92	24 (*R*)
3	$Mg(OTf)_2$	85	2 (*R*)
4	$MgBr_2$	97	52 (*R*)

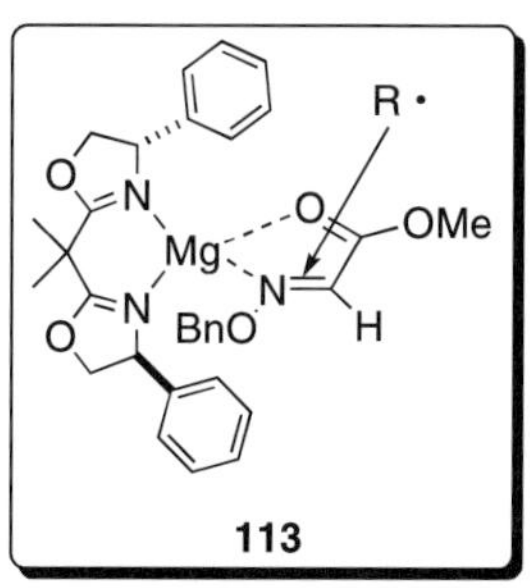

Scheme 29 Addition to glyoxylate imines

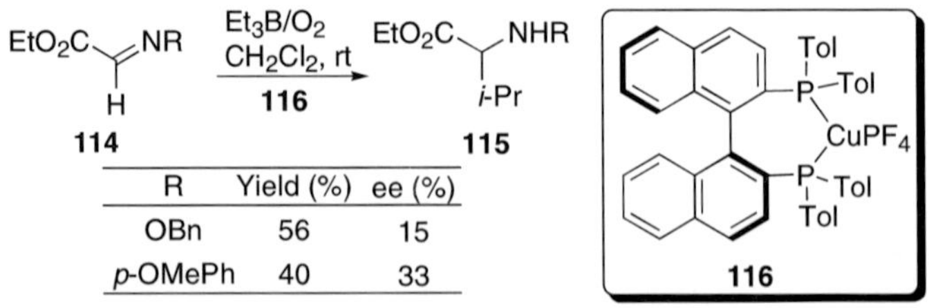

R	Yield (%)	ee (%)
OBn	56	15
p-OMePh	40	33

Scheme 30 Addition to glyoxylate imine using Cu-BINAP

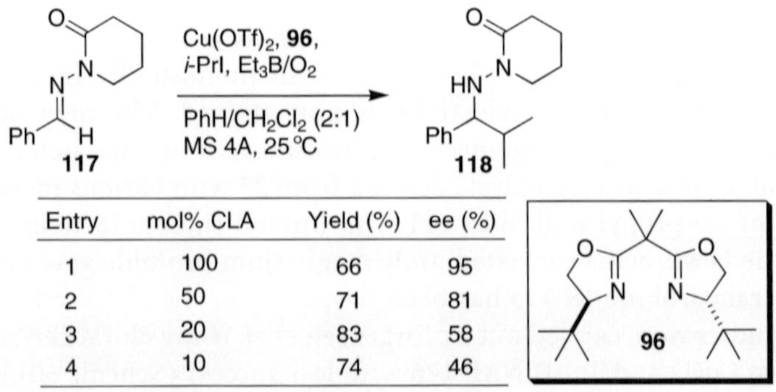

Entry	mol% CLA	Yield (%)	ee (%)
1	100	66	95
2	50	71	81
3	20	83	58
4	10	74	46

Scheme 31 Enantioselective radical additions onto hydrazones

Friestad and co-workers recently demonstrated that *N*-acyl hydrazones were excellent radical acceptors in the presence of a chiral Lewis acid [84]. Valerolactam-derived hydrazone **117** proved to be the optimal substrate for enantioselective radical additions. Upon further optimization it was found that $Cu(OTf)_2$ and *t*-butyl bisoxazoline ligand **96** gave the best yields and ee's (Scheme 31). Interestingly, a mixed solvent system (benzene:dichloromethane in a 2 : 1 ratio, respectively) in the presence of molecular sieves (4 Å) were necessary to achieve high yields and selectivities.

3.2 Conjugate Addition

Intermolecular conjugate addition [85] of nucleophilic radicals to α,β-unsaturated compounds has been carried out enantioselectively using chiral Lewis acids. Sibi, Porter and co-workers showed that magnesium and zinc Lewis acids along with bisoxazoline ligands can catalyze the reaction of oxazolidinone cinnamate **119** with *i*-PrI to give addition product **120** (Scheme 32) [86]. The success of this process depends on the activation provided by the Lewis acids, which make the reaction possible at – 78 °C. The non-Lewis acid mediated process is negligible at this temperature. Bidentate chelation of the substrate and chiral ligand to the Lewis acid generates the

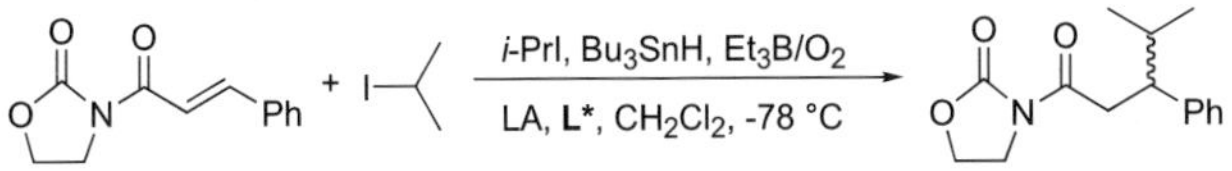

119 122 120

Entry	LA	L* (eq)	Yield (%)	ee (%)
1	MgI_2	**121** (1.0)	88	82 (*R*)
2	MgI_2	**121** (0.5)	86	79 (*R*)
3	MgI_2	**93** (1.0)	88	47 (*S*)
4	$Zn(OTf)_2$	**93** (1.0)	88	61 (*S*)
5	MgI_2	**122** (1.0)	88	93 (*R*)
6	MgI_2	**122** (0.3)	91	97 (*R*)
7	MgI_2	**122** (0.1)	88	95 (*R*)
8	$Zn(OTf)_2$	**122** (1.0)	-	53 (*R*)

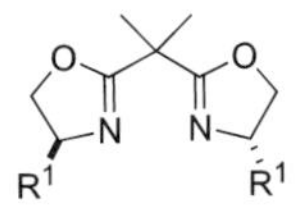

121 R^1 = *i*-Bu
93 R^1 = Ph

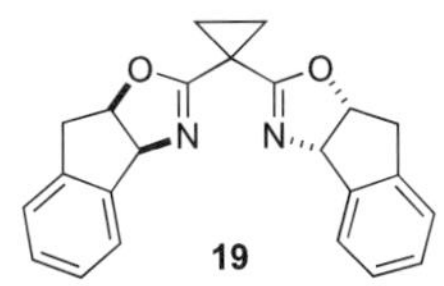

19

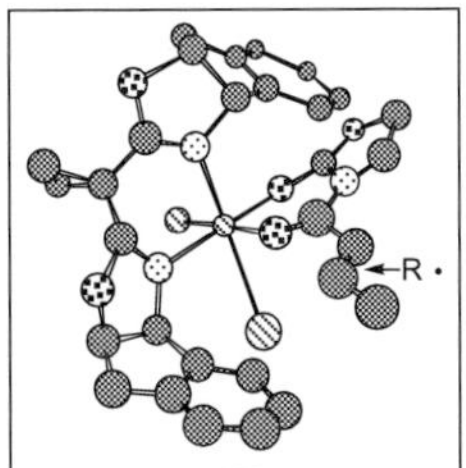

123

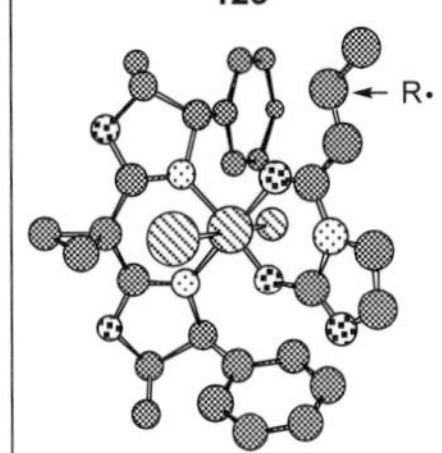

124

Scheme 32 Conjugate addition to oxazolidinone cinnamate

reactive complex. The substrate adopts a *s-cis* orientation at the C(O)-C(sp^2) bond when bound to the Lewis acid. The use of Et_3B, an efficient initiator at low temperatures, in the presence of oxygen generates radicals from haloalkanes. These radicals then add to the substrates bound to the chiral Lewis acid in an enantioselective manner. Interestingly, face selection depends on whether the C-4 substituent on the bisoxazoline ligand has an alkyl **121** or an aryl **93** group. Moreover, zinc Lewis acids gave good selectivities with **93** whereas magnesium salts gave good selectivities with **19**. The process was shown to be catalytic in the chiral Lewis acid. Sibi and Ji then evaluated various bisoxazoline ligands with MgI_2 and found that ligands derived from cis-aminoindanol were more effective in these reactions [87]. Further optimization based on the ring size at the bridging carbon showed ligand **19** to be the best ligand. This combination of MgI_2 and **19** catalyzed the reaction even at 10 mol % loading (entry 3) and also at room temperature (entry 5) without significant loss in enantioselectivity. The observed stereochemical outcome of the reactions was explained using octahedral models as shown in **123** and **124**. A cis-octahedral model **123** is proposed with **19** whereas a trans-octahedral model **124** accounts for selectivity with **93**. A more recent report of enantioselective conjugate radical additions onto **119** by Sibi and Petrovic displays the potential of metal triflimides as chiral Lewis acids [88].

Curran and Kanemasa explored the use of DBFOX/Ph ligand in conjugate radical additions [89]. This ligand had previously proven to be effective in Diels–Alder reactions. Evaluation of various main group and transition metal Lewis acids revealed that only $Mg(ClO_4)_2$ gave good reactivity (100% yield) and enantioselectivity (75% ee) (data not shown). DBFOX, a tridentate ligand, increases the electron density on Mg and makes it a weaker Lewis acid. This leads to the non-selective background reaction (non Lewis acid catalyzed) and hence to the lowering of enantioselectivity.

Murakata et al. have also examined enantioselective conjugate addition as shown in Scheme 33 [90]. In an effort to evaluate the role of additives in chiral Lewis acid mediated reactions they chose $Zn(OTf)_2$ as the Lewis acid. Ligand **128** with diethyl substitution at the bridging carbon of bisoxazoline was utilized throughout this study. Under stoichiometric chiral Lewis acid, very low enantioselectivity of the product was observed. When additives **129** and **130** were added there was a marked increase in ee's (entries 2 and 3). This increase was more dramatic when 4,4-diphenyl substituted oxazolidinone **125** was used as the template along with **130** as the additive (entries 4 and 5). Additional experiments (including low temperature NMR experiments) were conducted which suggested that the additives **129** and **130** were coordinating to zinc through the NH group. It was also possible for the reaction to be carried out at substoichiometric loading of chiral Lewis acid without substantial loss in selectivity (entry 6).

The templates used in these reactions have a significant impact on the reaction outcome and in determining product enantioselectivity. Sibi et al. also showed that changing the oxazolidinone template as in **119** to a 3,5-dimethylpyrazole in **131** resulted in a reversal of stereochemistry using the same chiral Lewis acid (Scheme 34) [91]. Additions in the presence of stoichiometric amounts of zinc triflate and ligand **19** gave good yields and moderate selectivities of **132**. These acylated pyrazoles **131** form 5-membered

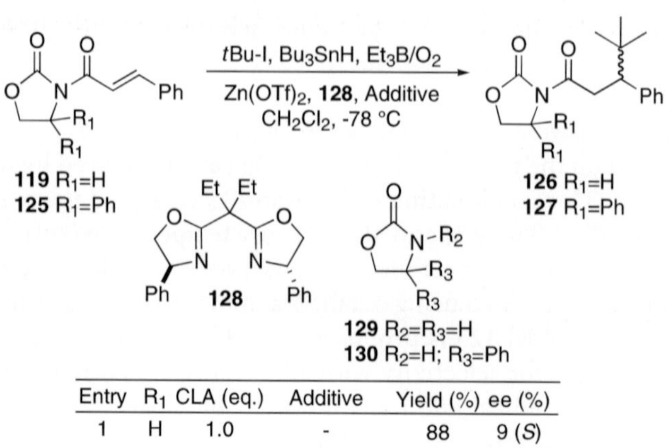

Entry	R_1	CLA (eq.)	Additive	Yield (%)	ee (%)
1	H	1.0	-	88	9 (*S*)
2	H	1.0	**129**	86	41 (*S*)
3	H	1.0	**130**	78	52 (*S*)
4	Ph	1.0	-	80	3 (*S*)
5	Ph	1.0	**129**	96	88 (*R*)
6	Ph	0.25	**130**	72	83 (*R*)

Scheme 33 Oxazolidinone additives in $Zn(OTf)_2$ catalyzed conjugate additions

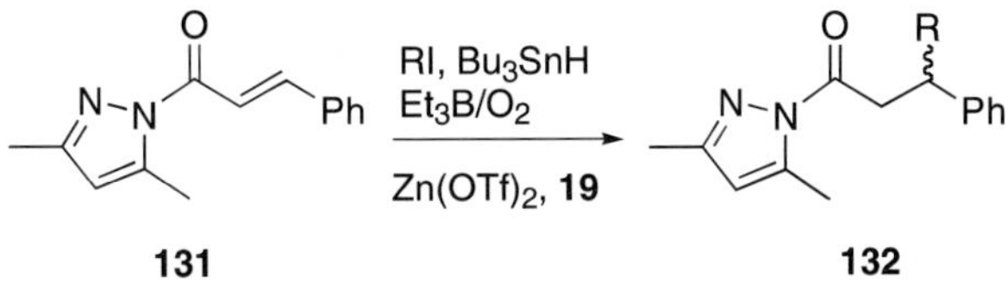

R	Yield (%)	ee (%)
i-Pr	76	51 (*S*)
Et	80	39 (*R*)
c-C_6H_{11}	81	81

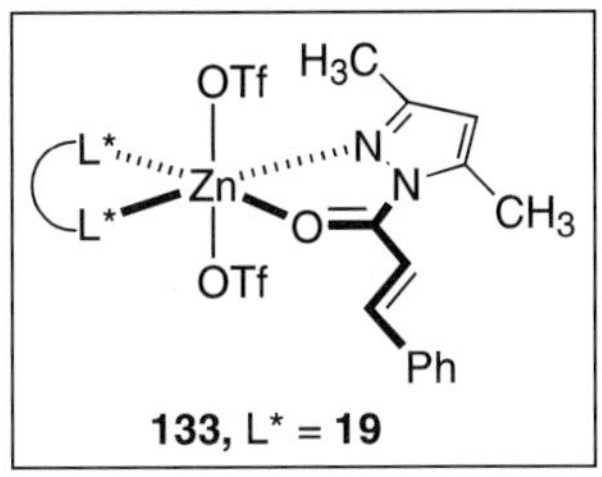

Scheme 34 Pyrazole templates: 5-membered chelation for reversal of enantioselectivity

chelates unlike the six-membered chelate formed with oxazolidinones **119** (vide supra). This change in chelate ring size, accompanied by a *trans*-octahedral geometry with **131** and **19**, has been proposed to account for the reversal of enantioselectivity (see **133**).

Another approach in attaining excellent enantiocontrol in conjugate radical additions is the development of novel achiral templates. Sibi and co-workers have developed an interesting template which contains a nitrogen fluxional group that can work in conjunction with a chiral ligand (Scheme 35) [92]. Using a small chiral bisoxazoline ligand (**136**), excellent ee's for product **135** can be obtained when the fluxional nitrogen substituent is

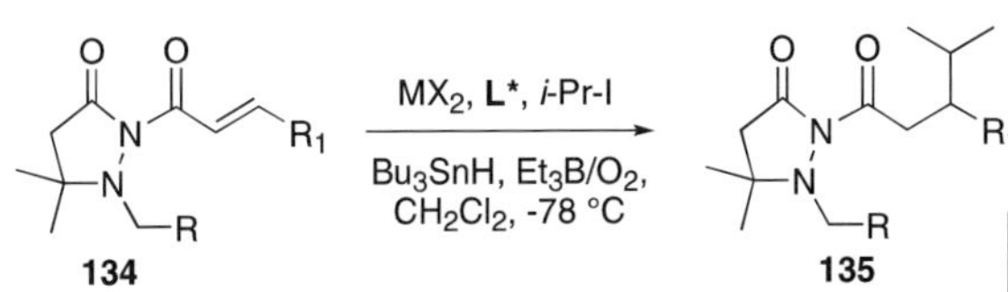

Entry	L*	MX_2	R	R_1	Yield (%)	ee (%)
1	**136**	$Cu(OTf)_2$	*n*-Pr	Ph	95	52
2	**136**	$Cu(OTf)_2$	Ph	Ph	92	82
3	**136**	$Cu(OTf)_2$	1-Nap	Ph	91	90
4	**19**	MgI_2	Ph	Ph	95	99
5	**137**	$Cu(OTf)_2$	Ph	Ph	94	-98
6	**19**	MgI_2	Ph	Me	85	98
7	**137**	$Cu(OTf)_2$	Ph	Me	80	-95
8	**19**	MgI_2	Ph	CO_2Et	50	56
9	**137**	$Cu(OTf)_2$	Ph	CO_2Et	52	-58

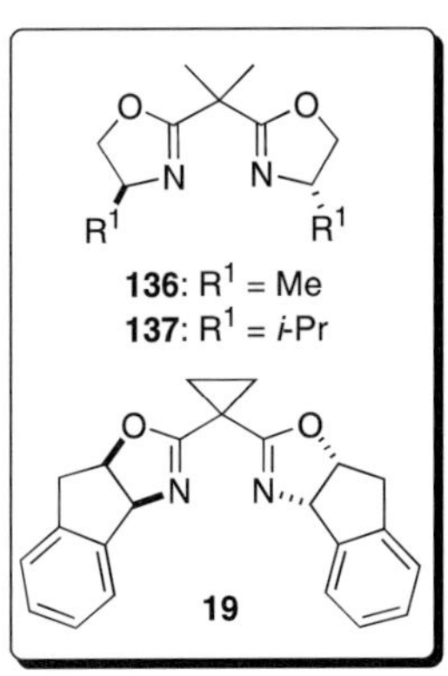

Scheme 35 Fluxional template for radical enantioselective conjugate additions

large (see entries 1–3). Interestingly, by selecting the proper Lewis acid it was found that enantiomeric products could be produced in near optically pure form (compare entries 4 and 5; 6 and 7). This is due to the fact that copper(II) triflate prefers a square planar geometry while magnesium Lewis acids adopt a tetrahedral or octahedral environment.

All examples of conjugate additions outlined above have utilized either main group or transition metal Lewis acids. In order to expand the scope of these reactions Sibi and Manyem developed a lanthanide Lewis acid-ligand system (Scheme 36) [93]. Lanthanide Lewis acids are unique in that they are less sensitive to air and moisture (ease of handling) and also make it possible for reactions to be carried out in aqueous media [94]. After a brief survey of lanthanide triflates, samarium triflate in the presence of ligand **138** was found to be the best combination with 30 mol % of catalytic loading being optimal. Evaluation of various substitutions in the ligand allowed determination of the importance of different groups. The aryl groups in the tertiary alcohol were necessary for good selectivity. After determining that the product was binding to the chiral Lewis acid and lowering the ee's (see entry 2), the importance of additives in improving selectivity was exemplified. Among various additives, acyloxazolidinone **139** was the best and two equivalents relative to the chiral Lewis acid was required. A size dependence of the substituent on the exo carbonyl of the additive was also investigated (not shown). Two equivalents of benzoyl oxazolidinone **139** along with MS 4 Å in addition to chiral Lewis acid gave the highest selectivity (entry 4). The authors proposed that the additives aid in blocking the vacant coordination sites in the lanthanide complex and hence making a more robust complex.

A recent application of enantioselective conjugate radical additions was seen in the synthesis of (+)-ricciocarpins A and B [95]. The key step in the synthesis was an asymmetric addition of a functionalized radical precursor **141** to afford intermediate **142** (Scheme 37). A chiral catalyst screening revealed that MgI_2 and bisoxazoline ligand **19** was optimal for achieving

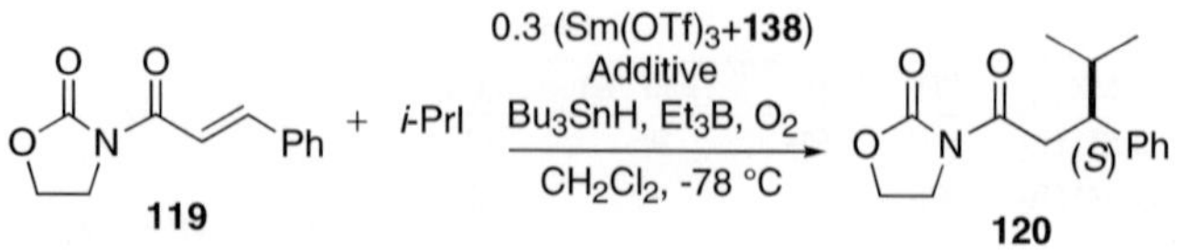

Entry	Conditions	Yield (%)	ee (%)
1	-	84	79
2	(±) product	83	49[a]
3	**139** (2eq)	67	89
4	**139** (2 eq)+MS 4Ă	63	92

[a] corrected for added (±) **120**

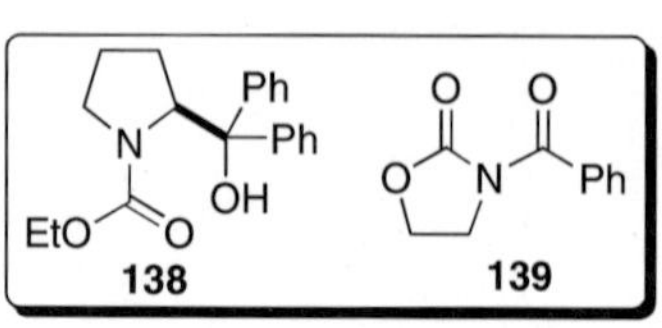

Scheme 36 Lanthanide mediated conjugate addition

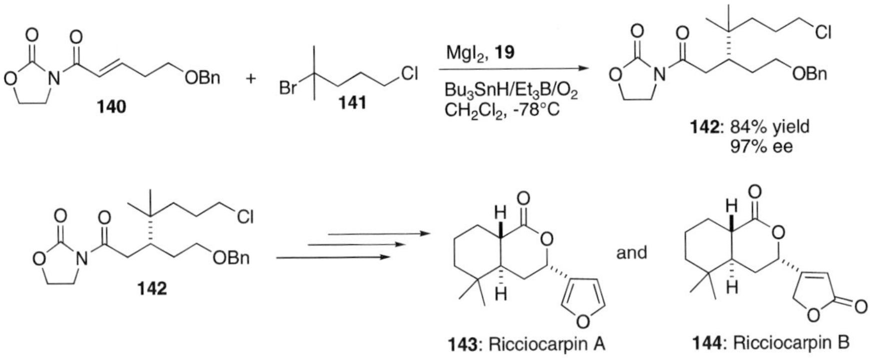

Scheme 37 Application of enantioselective conjugate radical addition: total synthesis of Ricciocarpin A and B

excellent yields and ee. A significant drop in conversion was observed by lowering the catalyst loading from 1.0 equivalent (84% yield, 97% ee) to 0.3 equivalent (16% yield, 93% ee). The synthesis of the targets ricciocarpins A and B were successfully completed in six and seven steps, respectively, from **140** (ricciocarpin A: 41% overall yield; ricciocarpin B: 45% overall yield).

Free radical chemistry provides an opportunity to explore reactions without the need for protecting groups on reactive functional groups such as alcohols. Recently, Sibi and Guerrero explored an enantioselective conjugate addition of haloalcohols to produce a variety of functionalized lactones [96]. Scheme 38 details the results for the formation of 6- and 7-membered lactones. A screening of an array of achiral templates revealed that 3,5-dimethyl pyrazole was optimal for both yield and enantioselectivity. Using substrates **145** in the presence of $Mg(NTf_2)_2$ and ligand *ent*-**19** a variety of lactones can be prepared in a single synthetic operation.

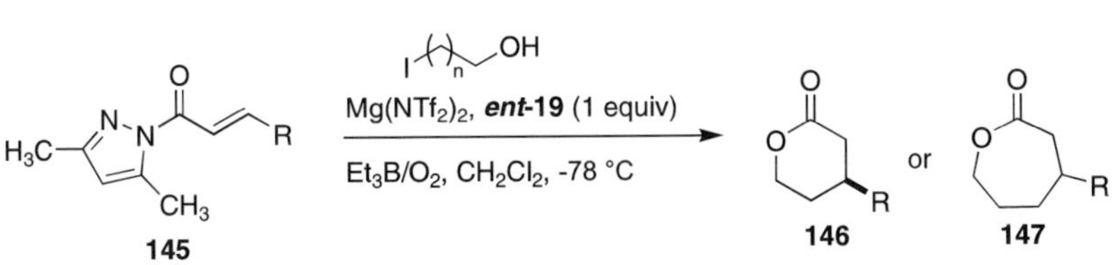

Entry	R	n	Prod.	Yield (%)	ee (%)
1	Ph	1	**146**	48	85
2	$PhCH_2CH_2$	1	**146**	50	85
3	$PhCH_2CH_2$	2	**147**	68	92
4	4-CF_3Ph	1	**146**	81	80
5	4-CF_3Ph	2	**147**	64	87

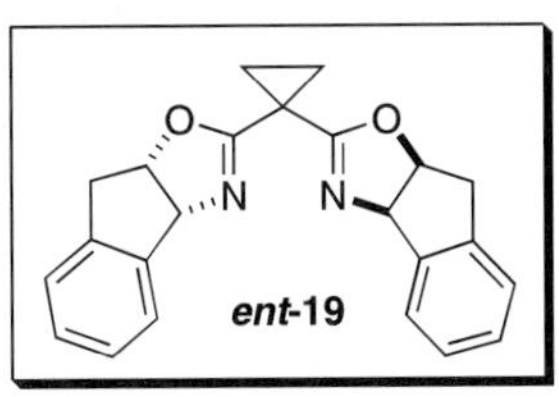

Scheme 38 Enantioselective lactone formation via haloalcohol radical conjugate addition

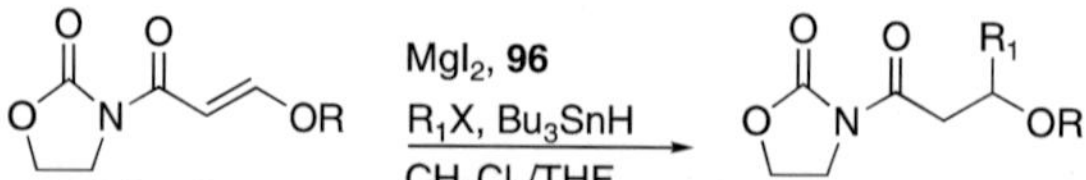

148 R = Bz
149 R = Ac

150

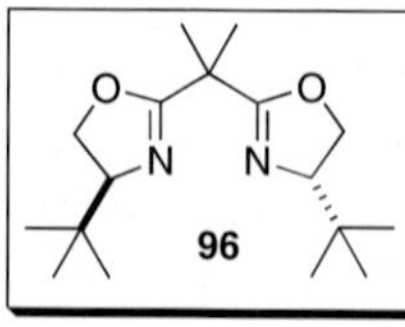

96

			30 mol% CLA		100 mol% CLA	
Entry	R	R_1	Yield (%)	ee (%)	Yield (%)	ee (%)
1	Bz	Et	82	33	90	50
2	Bz	*i*-Pr	83	58 (*R*)	90	93 (*R*)
3	Bz	*t*-Bu	91	46 (*R*)	91	89 (*R*)
4	Ac	*t*-Bu	70	30	73	95

Scheme 39 Acetate aldols by enantioselective conjugate radical additions

The aldol reaction is one of the most important reactions in synthetic organic chemistry. Many traditional ionic routes are currently available for diastereo- and enantioselective aldol reaction [97–99]. In contrast to highly basic ionic processes, development of radical methods for preparation of aldols using neutral conditions is attractive [100–102]. With the exception of intramolecular cyclization reactions, radical approaches towards aldol products remain largely unexplored [103–109].

An investigation of the acetate aldol-like radical reaction is shown in Scheme 39 [36]. It was discovered that benzoate **148** was the optimal substrate since it had shown the best characteristics with respect to yield and selectivity. Magnesium iodide in combination with ligand **96** was used as the chiral Lewis acid. A brief study on catalyst stoichiometry using **148**, **96**, MgI_2, and *i*-Pr – I showed a steady increase in ee with loading (50 mol % catalyst, 59% ee; 70 mol % catalyst, 80% ee; 90 mol % catalyst, 85% ee) and reaching a maximum with one equivalent (93% ee). These results suggest that either the catalyst turnover is slow[1] or that non-catalyzed reaction competes with the catalyzed process. In general, different types of radicals were chemically efficient irrespective of their nature (primary, secondary, or tertiary) or size with yields ranging from 70–90% (entries 1–3). On the other hand, the enantioselectivity varied to some extent depending on the radical precursor. Addition of a primary ethyl radical gave **150** in moderate ee at both 30 and 100 mol % catalyst loading (entry 1). Acyclic secondary radicals gave excellent selectivity (93% ee) in the conjugate addition (entry 2). Bulky tertiary radicals were equally effective in the conjugate addition. The addition of *t*-butyl radical gave good selectivity (entry 3). To investigate if the size of the acyloxy group impacted selectivity, tertiary radical addition to **149**, the acetate, was under-

[1] The aldol product containing donor atoms may not readily dissociate from the chiral Lewis acid and thus compete for coordination with the substrate. This explanation is consistent with the need for stoichiometric amounts of the chiral Lewis acid to obtain high ee's. REACT IR studies provide additional support for our explanation. These results will be reported in a full account.

taken. There was a significant improvement in selectivity for *t*-butyl radical addition (compare entry 3 with 4).

The absolute stereochemistry for **150** (entries 2 and 3) was determined by hydrolysis and conversion to known compounds. Assuming a tetrahedral or cis octahedral geometry for the magnesium [110], the product stereochemistry is consistent with *si* face radical addition to an *s-cis* conformer of the substrate. This is the same sense of selectivity as that obtained with oxazolidinone crotonates or cinnamates suggesting that the rotamer geometry of the differentially substituted enoates is the same. The need for stoichiometric amount of the chiral Lewis acid to obtain high selectivity with **148** in contrast to successful catalytic reactions with crotonates is most likely a reflection of the additional donor atom present in the substrate.

There are few addition reactions to α,β-disubstituted enoyl systems **151** that proceed in good yield and are able to control the absolute and relative stereochemistry of both new stereocenters. This is a consequence of problematic $A^{1,3}$ interactions in either rotamer when traditional templates such as oxazolidinone are used; to relieve $A^{1,3}$ strain the C – C bond of the enoyl group twists, breaking conjugation which results in diminished reactivity and selectivity [111–124]. Sibi et al. recently demonstrated that intermolecular radical addition to α,β-disubstituted substrates followed by hydrogen atom transfer proceeds with high diastereo- and enantioselectivity (**151** → **152** or **153**, Scheme 40).

The results obtained from the addition of various radicals R_1 to substrate **154** is shown in Scheme 41 [125]. Addition of a primary ethyl radical gave moderate yield and reduced diastereoselectivity, but the enantioselectivity for the major *anti* product was excellent (entry 1). Addition/trapping with the secondary isopropyl radical gave excellent yield of the *anti* product in high enantioselectivity (entry 2). *Tert*-butyl radical gave the *anti* isomer with outstanding selectivity (entry 5). The impact of changing the α- and β-substituents on the substrate is shown in entries 4–6. A decrease in yield, diastereo- and enantioselectivity was observed on changing the β-substituent R_2 from a methyl group to an ethyl group (compare entry 3 with 4). However, changing the β-substituent to a phenyl group gave the addition/trapping product with very high selectivity (entry 5). A larger α-ethyl substituent was less well tolerated leading to reduced selectivity (entry 6).

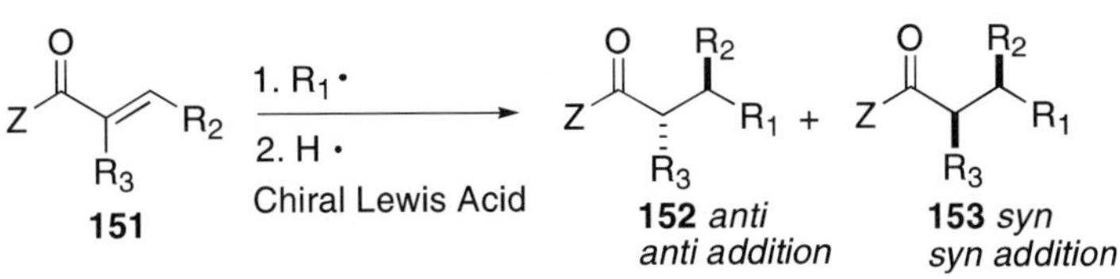

Scheme 40 Radical addition to α,β-disubstituted substrates

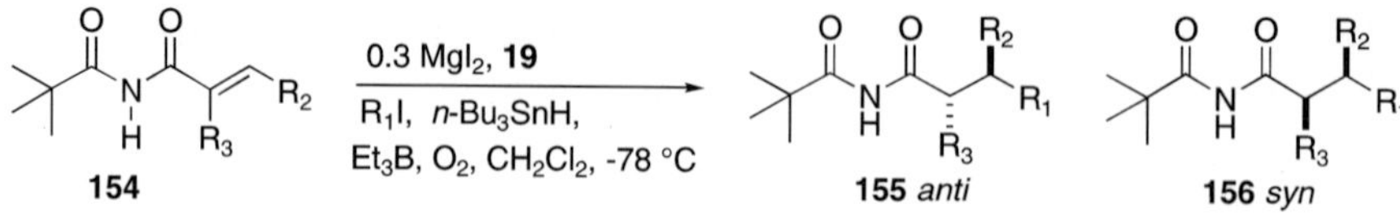

Entry	R_1	R_2	R_3	Yield (%)	dr	ee (%)
1	Et	Me	Me	63	87:13	92
2	*i*-Pr	Me	Me	79	99:1	92
3	*t*-Bu	Me	Me	83	99:1	94
4	*i*-Pr	Et	Me	37	95:5	80
5	*i*-Pr	Ph	Me	71	99:1	93
6	*i*-Pr	Me	Et	50	87:13	74

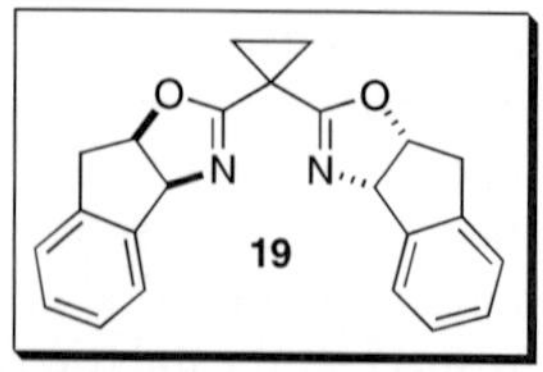

Scheme 41 Enantioselective radical additions to tiglate substrates

There are many routes for the synthesis of *syn*-aldols, yet the number of highly selective methods for preparing *anti*-aldols is limited [126–132]. Sibi and coworkers have recently shown that acetate aldols are accessible through enantioselective conjugate radical additions to β-acyloxyenoyl oxazolidinones [36]. Initial attempts to add radicals to α-methyl-β-acyloxy oxazolidinones, however, gave negligible reactivity (< 10%); but greatly improved reactivity results when an N – H imide template lacking the $A^{1,3}$ strain is used, making possible a highly diastereo- and enantioselective method for the preparation of *anti*-propionate aldol-like products **158** (Scheme 42). $Mg(ClO_4)_2$-**19** was used as the chiral Lewis acid and with all three radicals

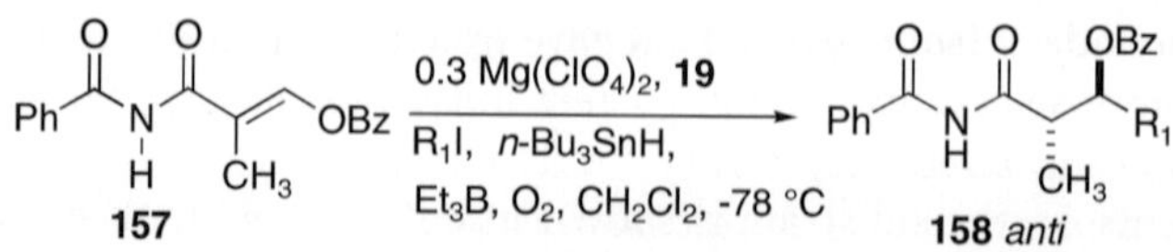

Entry	mol % CLA	R_1	Yield (%)	dr	ee (%)
1	30	*i*-Pr	79	99:1	82
2	30	*t*-Bu	60	99:1	90
3	30	*c*-hexyl	64	99:1	70

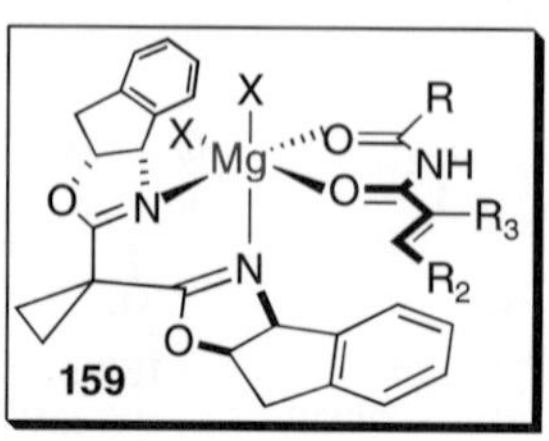

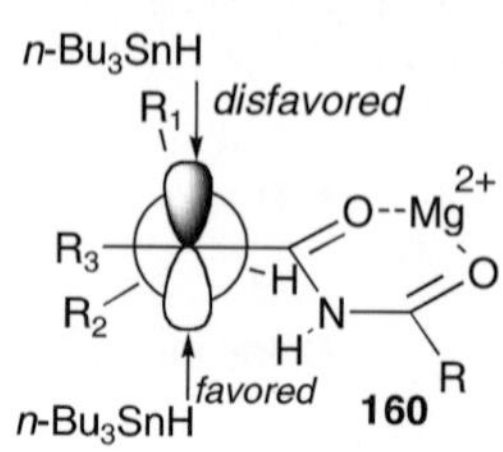

Scheme 42 Propionate aldols by enantioselective conjugate radial additions

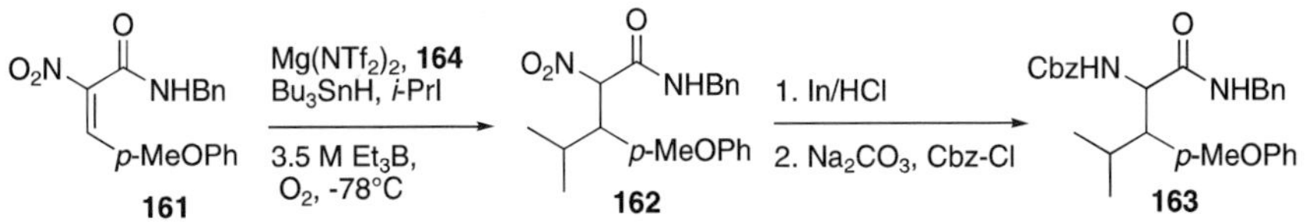

Entry	mol% CLA	Yield (%)	*syn* / *anti*	ee (%) (*syn*, *anti*)
1	100	76	1.4:1	88, 76
2	50	70	1.4:1	84, 64
3	30	57	1.2:1	41, 30

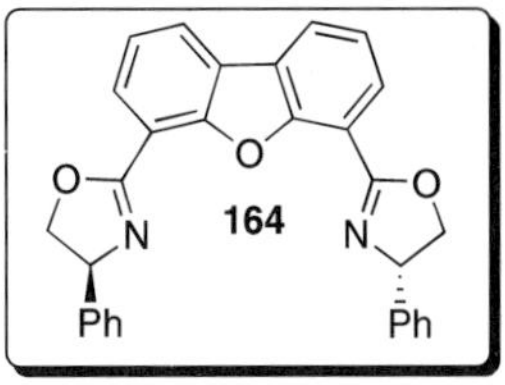

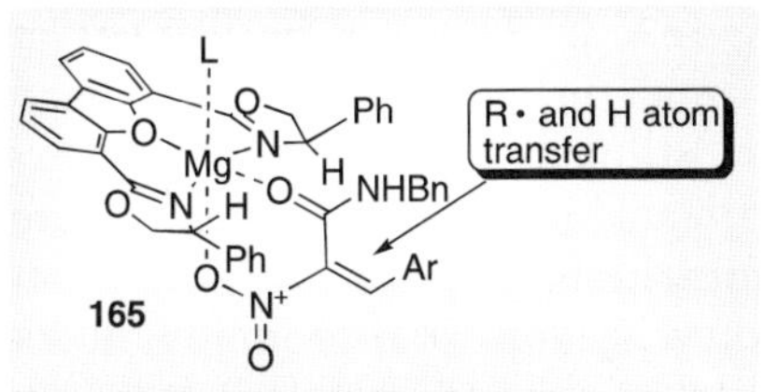

Scheme 43 β-Substituted α-amino acids via enantioselective radical conjugate additions

screened, yields are good, enantioselectivity is high, and the *anti* diastereoselectivity is outstanding.

A recent report by Castle and coworkers describes the synthesis of β-substituted α-amino acids through an enantioselective conjugate radical addition followed by H-atom transfer [133]. A screening of Lewis acids and chiral ligands revealed that a combination of $Mg(NTf_2)_2$ and DBFOX/Ph-**164** provided the best yields and enantioselectivities (Scheme 43). An interesting observation in this methodology was the dependence of triethyl borane concentration on enantioselectivity (1.0 M Et_3B gave 28% ee while 3.5 M Et_3B gave 77% ee under identical conditions). Under optimized conditions a maximum of 88% ee was achieved but the diastereoselectivity was quite low at 1.4 : 1 (entry 1). The absolute configuration for both diastereomers of **158** was determined by conversion to the corresponding amino acids [(2*R*, 3*S*) for the syn isomer and (2*R*, 3*R*) for the anti isomer]. These results lead the authors to conclude that the conjugate addition is relatively non-selective while the H-atom transfer step proceeds with good selectivity since the α-radical center is closer to the chiral space (see **165**).

3.3 Cyclizations

Enantioselective cyclizations by radical additions to olefins have been reported and a few of them have already been discussed in Sect. 2.2.1. Cyclizations were performed by Nishida et al. using chiral aluminum Lewis acid derived from Me_3Al and BINOLs (Scheme 44) [134]. Formation of a vinyl radical followed by a 5-*exo* or 6-*exo* (for $n = 1$ or 2) cyclization controlled by the chiral Lewis acid provides enantiofacial selection. For the carboxylic esters **166** the cyclizations were carried out at – 78 °C for the cyclopentane formation and at 0 °C for the cyclohexane formation. Lower yields of the six-membered ring products are because of the difficulty in 6-*exo* cyclizations. The use of one equiv. of either **168** or **169** provided cyclized products in low ee's with **169** performing slightly better. Using 4 equiv. of **169**, cyclized products (*R*)-**167a** and (*R*)-**163b** were obtained in 36 and 48% ee respectively. When the ester was replaced with the Weinreb amide, the cyclization proceeded smoothly to provide (*S*)-**167c** in 26% ee. The low ee's are due to the background reaction of the amide in the absence of complexed **169**. The importance of the carboxylic substituents is evidenced in this example. The esters upon complexation are oriented in an *s-trans* fashion whereas the Weinreb amides adopt an *s-cis* conformation.

Hiroi and co-workers performed cyclizations of *α*-bromo-*N*-allyl amides and sulfonamides **170a–c** with radical generation using triethylborane and trapping the cyclized radical with tinhydride (Scheme 45) [135]. The use of various Lewis acids was explored and titanium tetraisopropoxide emerged to be superior to either triethylaluminum or magnesium triflate. Among the substrates, less bulky substituents on nitrogen resulted in better reaction efficiency, with larger substituents like 2,4,6-triisopropylphenyl- and 1-naphthalenesulfonyl leading to reduced products along with recovered starting materials. The substrate **170c** with *p*-tolyl sulfonyl (tosyl) substituent was ideal in terms of both reaction efficiency and enantioselectivity. The products obtained possessed *trans* geometry at the newly formed C – C bond. The best selectivity was obtained with ligand **173** for the *trans* product **171c**.

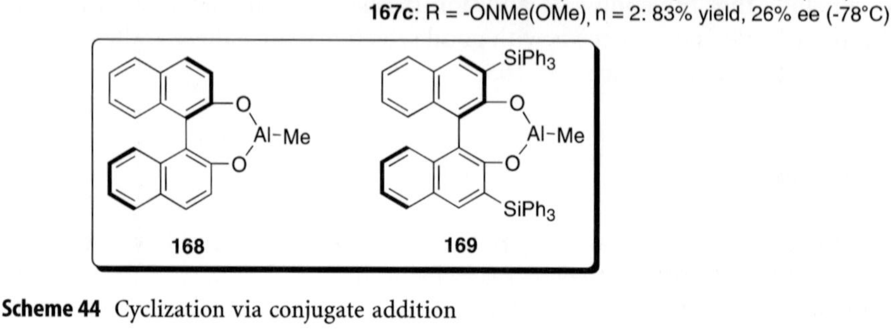

Scheme 44 Cyclization via conjugate addition

A unique cyclization procedure was conducted by Curran et al. in which they showed that axial chirality can be transferred into a new stereocenter with retention of chirality (Scheme 46) [136]. Substrates *M* or *P*-**174a,b**

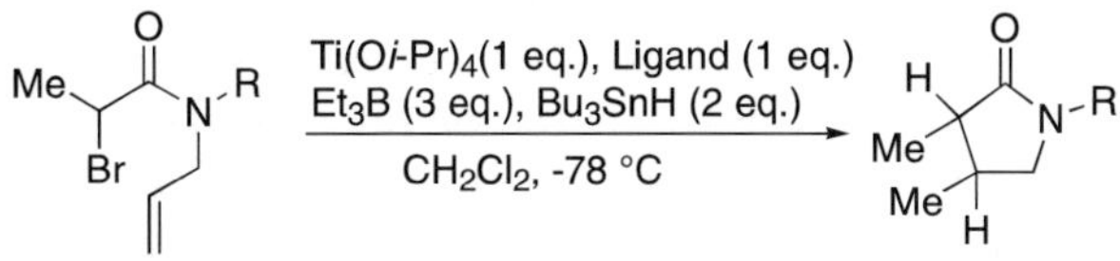

170a R = Ph
170b R = SO$_2$Me
170c R = Tosyl

171a
171b
171c

Entry	Substrate	Ligand	ee (%)
1	**170a**	**172**	5
2	**170b**	**172**	4
3	**170c**	**172**	47
4	**170c**	**173**	77

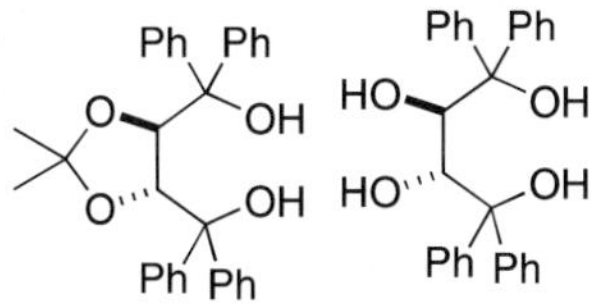

172 **173**

Scheme 45 Cyclization of α-bromo-*N*-allylamides and sulfonamides

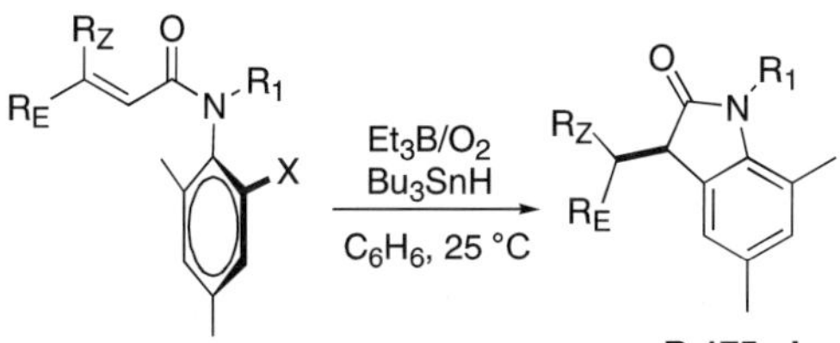

M-174a-b **R-175a-b**

Entry	Precursor	R_1	R_E	R_Z	X	Yield (%)	ee (%)
1	**M-174a**	Me	Me	H	I	73	89 (*R*)
2	**P-174a**	Me	Me	H	I	70	85 (*S*)
3	**M-174b**	Me	Ph	H	I	75	94 (*R*)
4	**P-174b**	Me	Ph	H	I	73	92 (*S*)

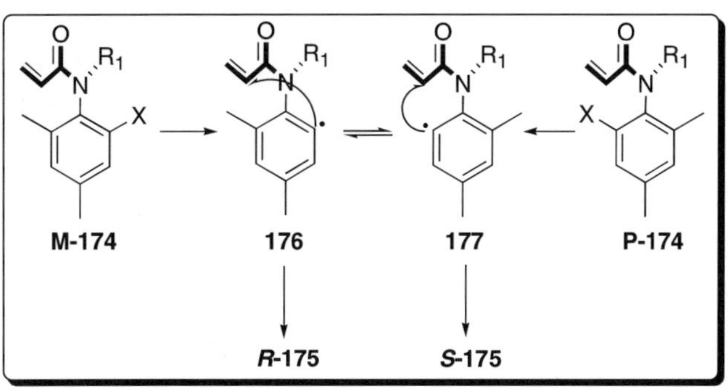

Scheme 46 Memory of chirality in cyclization

were either prepared from the chiral pool or by racemic synthesis followed by preparative chiral HPLC separation. When these were subjected to conditions shown in Scheme 46, the products (*R*) or (*S*)-**175** were obtained in good yields and high ee's. The high ee's are due to the almost complete absence of racemization of radical intermediates **176** or **177**. This is in turn related to the efficiency of the aryl radical addition to the olefin. The intermediate **176** obtained from *M*-**174** has to rotate around the aryl-nitrogen bond in order for the proper overlap required for cyclization to occur. If the cyclization is not efficient, there is a possibility of the bond rotation going further, leading to **177** and hence to racemization. These factors are borne out in the examples presented. It was observed that higher ee's are obtained when R_E is phenyl: the delocalization (and hence stabilization) provided by the carbonyl group becomes less important due to the delocalization provided by conjugation with the phenyl ring. This allows for *M*-**174c** to react faster furnishing higher ee's. A more recent report of this asymmetric cyclization of transient atropisomers has been reported by Curran and co-workers [137].

4 Fragmentations

Fragmentation reactions involve the addition of radicals to a neutral molecule followed by *β*-elimination of the resulting radical generating a terminal olefin [138]. The most common trap for a radical popularized by Keck is allyltributylstannane [139]. Porter et al. performed the trapping of acyl radicals obtained from *α*-bromo oxazolidinones **178** with allylstannanes and allylsilanes (Scheme 47) [140]. Magnesium and zinc Lewis acids were used with bisoxazolines to induce enantioselectivity. Although the allylstannanes are good trapping agents, this reaction produces stannyl halides that are Lewis acidic and can compete with the chiral Lewis acid in catalyzing the reac-

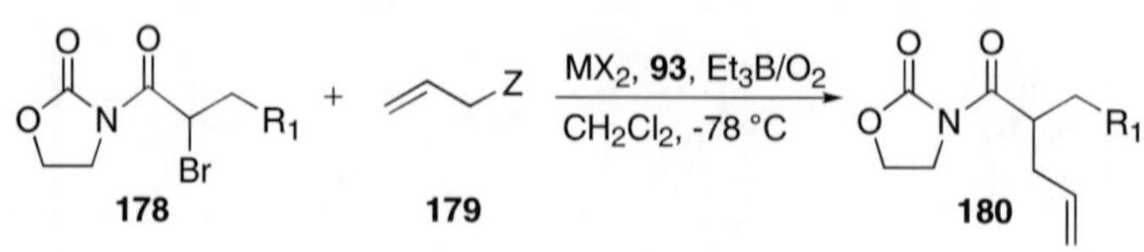

Entry	R_1	MX_2	Z	Yield (%)	ee (%)
1	Me	$Zn(OTf)_2$	$SnBu_3$	84	42 (*S*)
2	Me	$Zn(OTf)_2$	$Si(OEt)_3$	65	60 (*S*)
3	*t*-Bu	$Zn(OTf)_2$	$SnBu_3$	63	74 (*R*)
4	*t*-Bu	$Zn(OTf)_2$	$SiMe_3$	88	90 (*R*)
5	*t*-Bu	MgI_2	$SiMe_3$	86	68 (*S*)

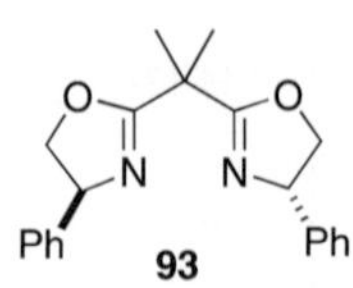

Scheme 47 Allylations of *α*-bromo oxazolidinones

tion. This results in racemic products, which then reduces the overall enantioselectivity of the products (entry 1). Replacement of allylstannanes with allylsilanes overcomes this problem and higher selectivities were obtained (compare entries 1 through 4). With the same bisoxazoline ligand, magnesium and zinc Lewis acids gave enantiomeric products (entries 4 and 5) as was observed in the conjugate addition reactions (*vide supra*). The authors also showed that added Me_3SnBr decreases enantioselectivity depending on the amount, supporting the hypothesis that this is the most probable cause for the decreased ee's in allylstannane reactions.

A similar study was published independently by Renaud and Fhal in which they reported the use of aluminum Lewis acids (Scheme 48) [141]. This study utilized the α-iodo-acyloxazolidinones **181a,b** as substrates and the corresponding radical was generated either by photolysis (– 10 °C) or using triethylborane (– 78 °C). The Et_3B method always gave better selectivities than photolysis. In the case of **181a**, (*S*)-BINOL **183** (**182a** = 0% ee) and the TADDOL **184** (entry 1) were ineffective. Substituents at C-4 can influence the *s-cis* versus *s-trans* conformers in the reactive radical species but in this instance, the increase in ee's for **181b** was less than expected (entries 2–4). Similar ee's were obtained with both allylstannane and methallylstannane (entries 3 and 4).

In a related study Porter et al. showed that α-bromo-γ lactams **185** containing a pyridyl moiety can react with allyltrimethylsilane enantioselectively in the presence of chiral Lewis acids derived from zinc and **189** (Scheme 49) [142]. In contrast to the above study, the ligand of choice for substrates **185** was found to be the bisoxazoline ligand **189**. Excellent ee's were obtained in the presence of two equivalents of the chiral Lewis acid. Under substoichiometric amounts of the catalyst, lower selectivities were obtained. Different substituents on the pyridyl moiety were also examined although no predictable trend was observed. A trans octahedral model simi-

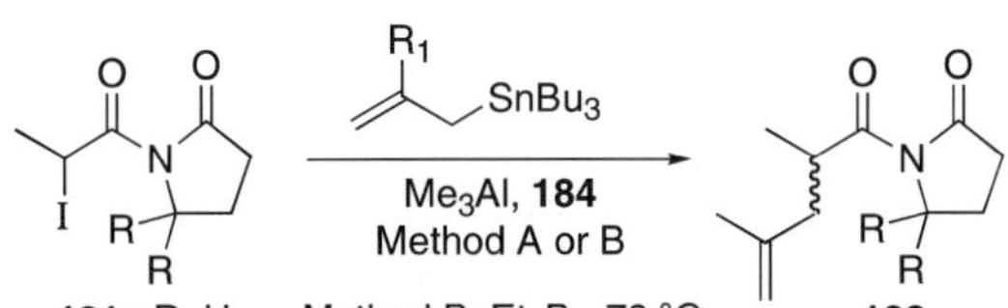

181a R=H
181b R=Me

Method B: Et_3B, -78 °C
Method A: AIBN/
300 W sunlamp, 10 °C

182

Entry	R	R_1	Method	Yield (%)	ee(%)
1	H	H	A	88	15 (*S*)
2	Me	Me	A	95	24 (*R*)
3	Me	Me	B	93	34 (*R*)
4	Me	H	B	90	32 (*R*)

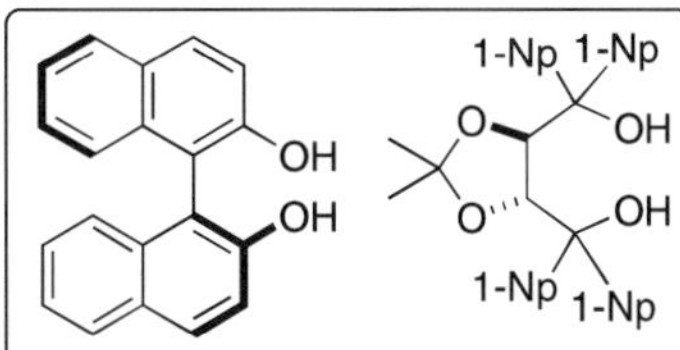

183 **184**

Scheme 48 Allylations mediated by trimethylaluminum

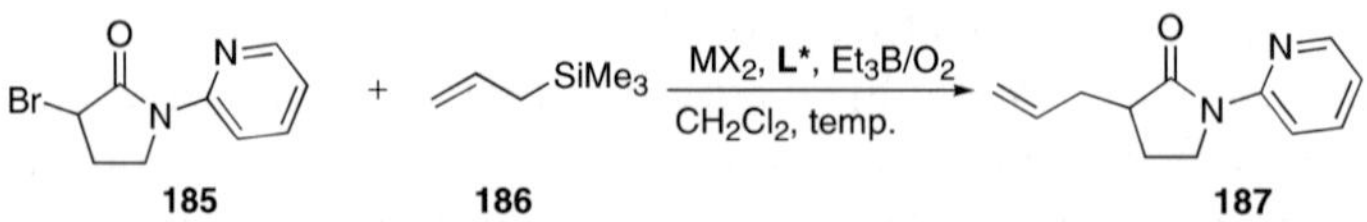

Entry	L*	MX2(eq.)	T (°C)	Yield (%)	ee (%)
1	**93**	$Zn(OTf)_2$ (1.0)	-78	70	59 (*S*)
2	**189**	$Zn(OTf)_2$ (1.0)	-78	75	96 (*S*)
3	**188**	$Zn(OTf)_2$ (1.0)	-78	42	75 (*R*)
4	**189**	$MgBr_2$ (1.0)	-78	54	84 (*R*)
5	**189**	$Zn(OTf)_2$ (2.0)	-78	83	>99 (*S*)
6	**189**	$Zn(OTf)_2$ (0.2)	-20	69	81 (*S*)

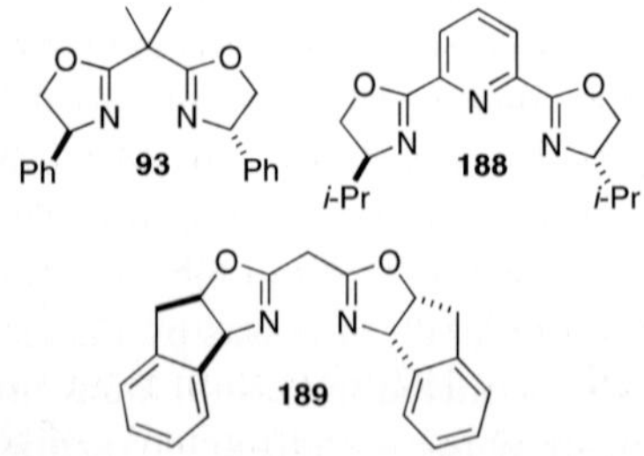

Scheme 49 Allylation of γ-lactams: pyridine as a template

lar to that in **124** was proposed to explain the stereochemical outcome of the reaction.

Hoshino and co-workers demonstrated the creation of chiral quaternary centers using allyltributylstannane [143]. In this study, a monodentate substrate **4** was used (Scheme 50; cf. Scheme 2). An initial evaluation of Lewis acids showed that among MgI_2, $MgBr_2$, $Zn(OTf)_2$, Et_2AlCl, and Me_3Al, only Me_3Al activated the substrate to provide the desired product. Use of this Lewis acid with ligand **191** provided the allylated product **190** in low (27%) ee. Addition of diethylether (1 equiv. to Lewis acid and ligand) pro-

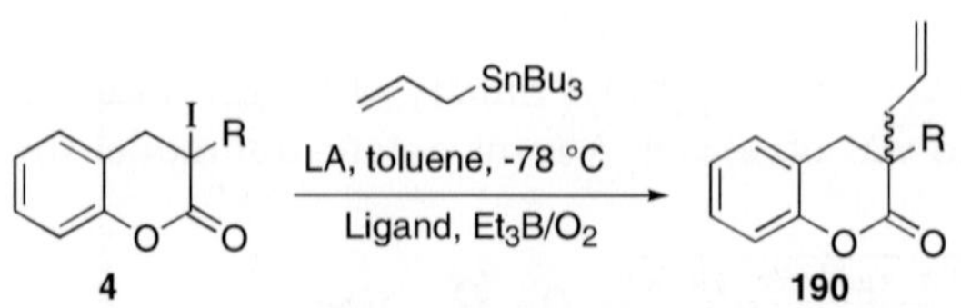

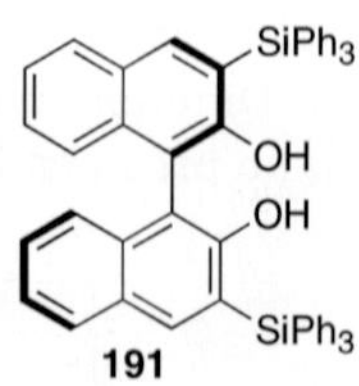

Entry	R	Ligand (eq.)	LA (eq.)	Additive	Yield (%)	ee (%)
1	Me	**191**(1)	Me_3Al (1)	-	72	27 (*R*)
2	Me	**191** (1)	Me_3Al (1)	Et_2O	84	81 (*R*)
3	CH_2OBn	**191** (1)	Me_3Al (1)	-	72	3 (*S*)
4	CH_2OBn	**191** (1)	Me_3Al (1)	Et_2O	76	91 (*R*)
5	CH_2OBn	**191** (0.2)	Me_3Al (0.2)	Et_2O	73	82 (*R*)
6	CH_2OBn	**192** (1)	Me_3Al (1.5)	-	73	54 (*R*)
7	CH_2OBn	**192** (1)	Me_3Al (2)	-	76	51 (*R*)
8	CH_2OBn	**192** (1)	*i*Bu_3Al (2)	-	75	43 (*R*)
9	CH_2OBn	**192** (0.5)	Me_3Al (1)	-	72	54 (*R*)

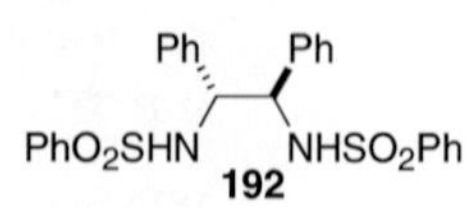

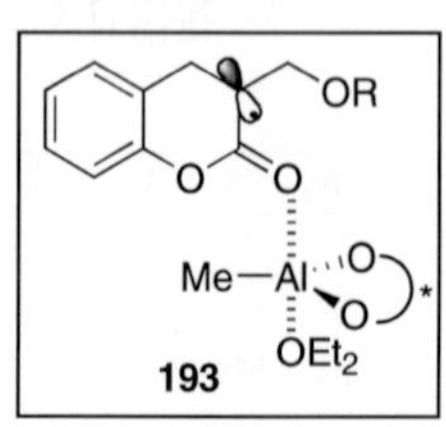

Scheme 50 Allylation of α-iodo lactones: Al-BINOL and Al-sulfonamide ligands as single point binding chiral Lewis acids

duced a dramatic increase in selectivity to 81% ee. Other additives such as diisopropyl ether, THF, or *N*-methylpyrrolidine (NMP) were not as effective as diethylether. Substrates containing either a simple methyl group or those with alkoxyalkyl group benefited from added ether (compare entries 2 and 4). The reaction was also shown to be catalytic in chiral Lewis acid (entry 5). A five-coordinate aluminum complex **193** has been proposed to account for the selectivity. Sulfonamido ligands derived from (*R*)-1,2-diphenylethylenediamine were also studied by Hoshino et al. with both trimethylaluminum and tri-isobutylaluminum with much less success: a maximum of 54% ee (Scheme 50) [144]. The chiral Lewis acids were prepared by refluxing **192** with aluminum salts followed by cooling to – 78 °C before the reaction. The heterogeneity of the catalyst in toluene was an issue. This was overcome by using larger amounts of the aluminum salts compared to the ligand or by the use of substoichiometric amounts of the catalyst. Similar levels of ee's (54%) were achieved in the lower loading of catalyst (entry 9).

In fragmentation reactions involving sulfonamides **194**, Hiroi and Ishii examined various Lewis acids with chiral diamines, diols, and sulfoxides (Scheme 51) [135]. Among these ligands, the sulfoxide ligand **198** gave good selectivity with magnesium triflate. The tosyl group in the substrate was important as the smaller methane sulfonamide gave much lower ee's (data not shown). The origin of selectivity with magnesium Lewis acid was explained using the model **199**. The orientation shown in **199** is the lowest energy conformation and the radical is trapped by the allylstannane from the face opposite to the *p*-tolyl group.

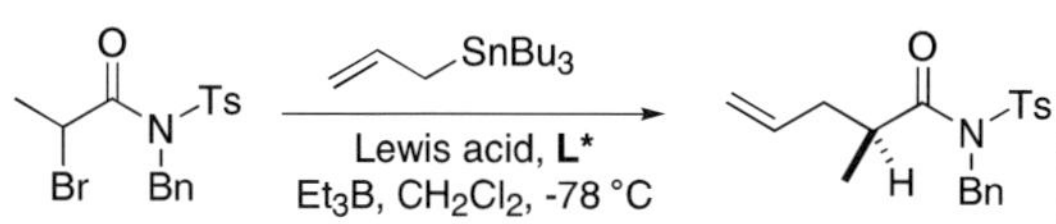

194 **195**

Entry	L*	Lewis acid	Yield (%)	ee (%)
1	**196a**	Et_3Al	88	53
2	**196b**	Et_3Al	80	64
3	**196c**	Et_3Al	83	63
4	**197**	$Mg(OTf)_2$	65	52
5	**198**	$Mg(OTf)_2$	41	83

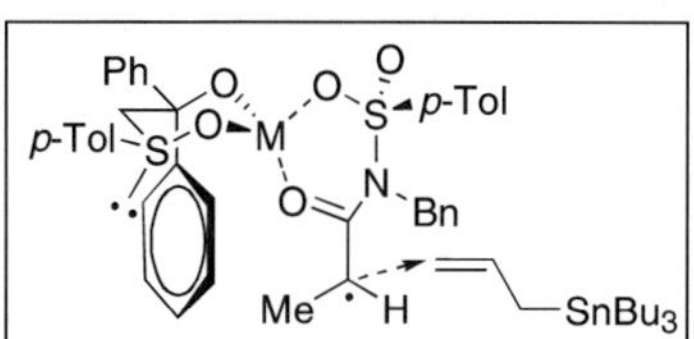

199

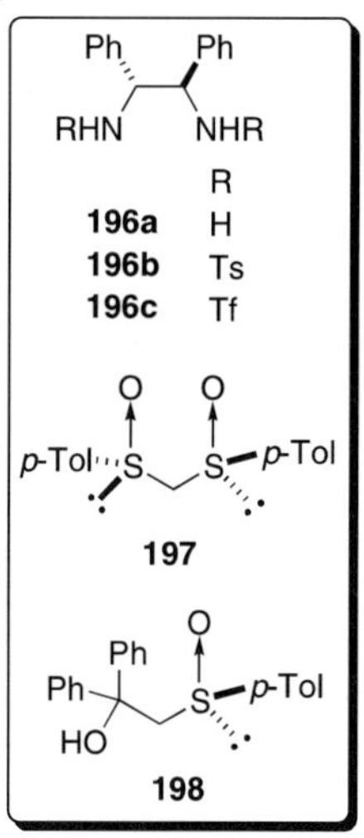

Scheme 51 Allylations using an acyclic template: sulfonamide

5 Tandem Reactions: Addition-Trapping

The previous section detailed the possibility of generating radicals followed by fragmentation reactions with allylstannanes. Such α-acyl radicals are intermediates in the conjugate addition of nucleophilic radicals to α,β-unsaturated compounds and can further react with allylstannanes. In doing so, a stereocenter is created at the carbon atom alpha to the carbonyl. In principle, one can create two stereocenters in this tandem reaction. A first example though, involved creation of a single chiral center: addition of radicals to oxazolidinone acrylate **200** followed by trapping with allylstannane (Scheme 52) [145]. Zinc triflate was found to be the ideal Lewis acid: zinc chloride, magnesium triflate, and scandium triflate were ineffective. Ether as solvent proved to be better than methylene chloride (entry 1 and 2): an anomalous behavior that possibly hints at a superior chiral catalyst with ether coordinated to zinc or stannane. Further experiments were carried out with CH_2Cl_2-pentane (40 : 60) mixtures and *t*-butyl radical additions gave up to 90% ee (entry 5). A tetrahedral zinc complex with bidentate chelation to substrate and ligand was proposed to account for the observed stereochemistry (**202**).

A direct comparison of the stereochemical efficiency of the fragmentation reaction versus the tandem reaction (Scheme 53) was studied by Porter et al. as a function of the steric effect based on the Taft parameters for different substituents [146]. In general, the tandem reactions perform better and provide higher levels of ee's than the fragmentation reactions. This effect could be due to the tinbromide by-product catalyzing a non-stereoselective process as has been uncovered by the same authors (*vide supra*) and by Sibi and Ji in their diastereoselective studies [147].

Sulfones are an appealing class of substrates and have been used in tandem reactions with generation of a chiral center α to the sulfonyl group (Scheme 54) [148]. In order to achieve bidentate chelation with metal, pyridyl

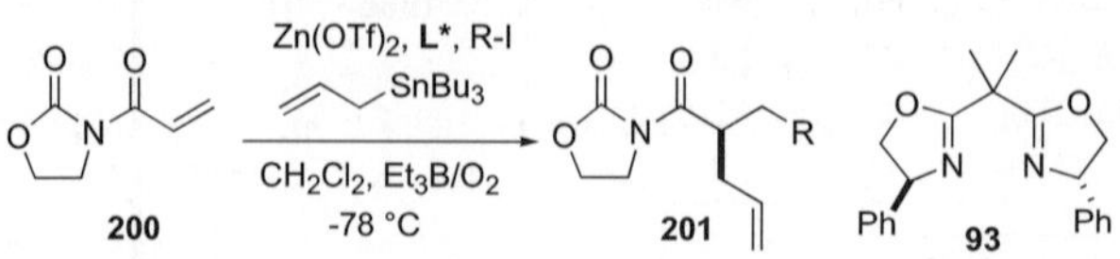

Entry	RI (eq.)	L*	**200**:M:L*	Solvent	Yield (%)	ee (%)
1	c-HexI (10)	(*S, S*)-**93**	1:1:1	CH_2Cl_2	62	50 (*S*)
2	c-HexI (10)	(*S, S*)-**93**	1:1:1	ether	90	64 (*S*)
3	c-HexI (10)	(*S, S*)-**93**	1:2:2	ether	61	80 (*S*)
4	c-HexI (1.5)	(*R, R*)-**93**	1:1:1.2	CH_2Cl_2:pent 4:6	62	68 (*R*)
5	*t*-BuI (5)	(*R, R*)-**93**	1:1:1.2	CH_2Cl_2:pent 4:6	92	90 (*R*)

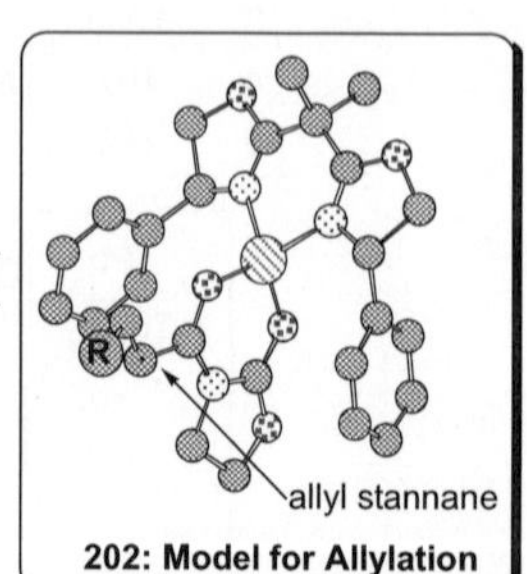

Scheme 52 Installation of α-stereocenter through addition-trapping

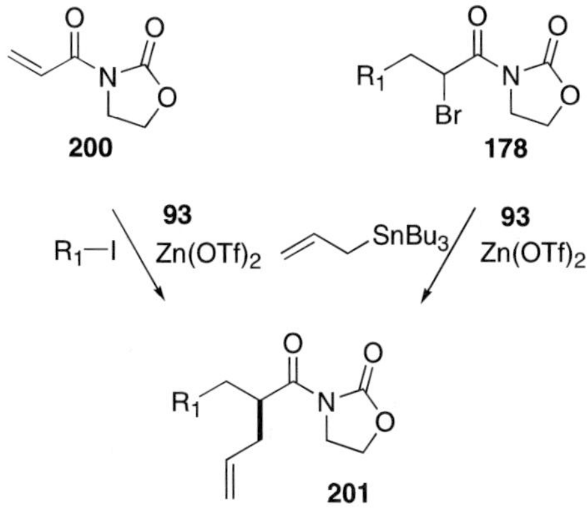

Scheme 53 Addition-trapping vs. fragmentation: influence of sterics on selectivity

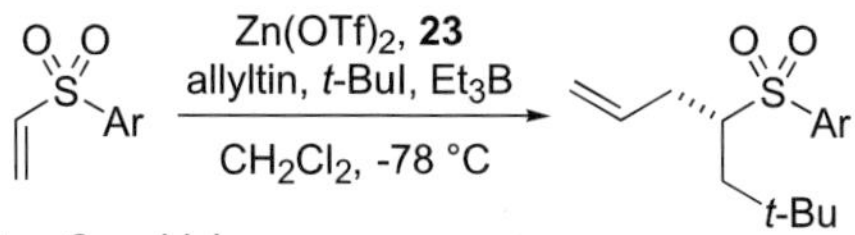

203a Ar = 2-pyridyl
203b Ar = 2-(1-benzylbenzimidazolyl)

(S)-**204**

Entry	Substrate	Allylstannane	Yield (%)	ee (%)
1	**203a**	$AllylBu_3Sn$	36	46
2	**203b**	$AllylBu_3Sn$	35	80
3	**203a**	$AllylPh_3Sn$	82	50
4	**203b**	$AllylPh_3Sn$	49	16
5	**203b**	$Allyl_4Sn$	87	51
6	**203b**	$Allyl_2Bu_2Sn$	74	50
7	**203b**	$Allyl_2Bu_2Sn$	86	78
8	**203b**	$Allyl_2Bu_2Sn$	80	84

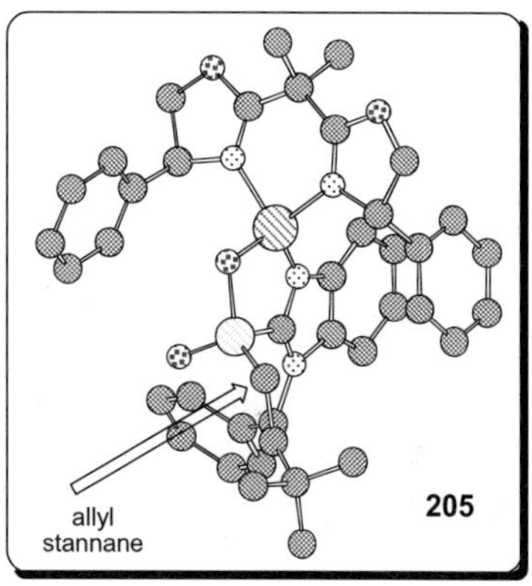

Scheme 54 Addition-trapping to sulfones

or benzimidazolyl moieties were also introduced in the substrates **203a,b**. Results indicate that the benzimidazolyl substrates **203b** perform better than pyridyl substrate **203a**. Diallyldibutylstannane performed better than all other stannanes. In addition, 10 equivalents of the stannane gave the best results in terms of yield and selectivity (entries 6, 7, and 8). A model proposed for the observed selectivity is shown in Scheme 53 (**205**).

Sibi and Chen demonstrated for the first time that relative and absolute stereocenters of both α and β carbons can be controlled in the intermolecular addition trapping experiments (Scheme 55) [149]. Magnesium and copper Lewis acids performed better than zinc. The use of 30 mol % of chiral Lewis acid gave higher selectivities than the stoichiometric amounts for both magnesium and copper. Interestingly, copper triflate gave better selectivities with

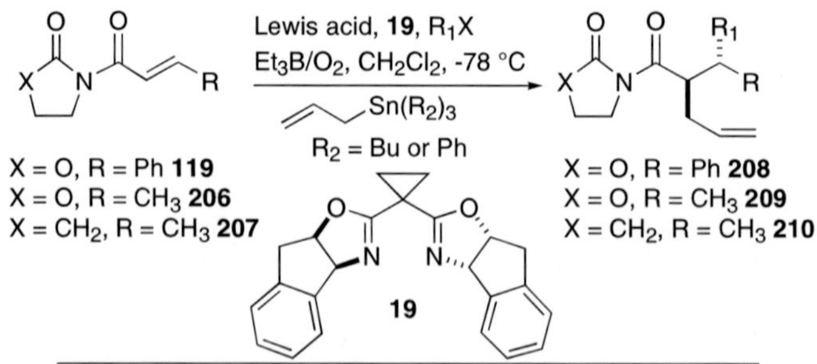

Entry	Substrate	LA (0.3 eq.)	R_1	Yield (%)	dr	ee (%)
1	**119**	MgI_2	Et	79	32	77
2	**119**	MgI_2	*i*-Pr	93	37	93
3	**119**	MgI_2	*t*-Bu	84	99	97
4[a]	**119**	$Cu(OTf)_2$	*t*-Bu	90	99	-96
5	**206**	$Mg(ClO_4)_2$	*c*-Hexyl	83	4	62
6	**207**	$Mg(ClO_4)_2$	*i*-Pr	84	7	76
7[a]	**207**	$Cu(OTf)_2$	*i*-Pr	95	10	-76
8	**207**	$Mg(ClO_4)_2$	*t*-Bu	85	19	92
9	**207**	$Cu(OTf)_2$	*t*-Bu	66	50	-83

[a] Allyltriphenyltin was used.

Scheme 55 Scope of addition-trapping reactions

allyltriphenyl stannane (entry 4) whereas no such difference was found with magnesium Lewis acids. Among the magnesium salts, iodide as the counter-ion was found to be more effective than bromide and perchlorate. The products were predominantly with *anti* stereochemistry. Another point of note was that copper triflate and magnesium iodide gave enantiomeric products (compare entries 3 and 4). A size dependence of the selectivity was noted with both the size of the radical being added or the size of the β-substituent. The best selectivities were obtained when *t*-butyl radical was added and for the β-phenyl substituent (entry 8). In the case of crotonate substrates **206** and **207** changing the template from oxazolidinone to pyrrolidinone led to higher selectivities. The *anti* selectivity was shown to mainly depend on the β-stereocenter.

A more recent report by Sibi and co-workers displayed the utility of chiral lanthanide Lewis acids for addition-trapping reactions [150]. An exhaustive screening of lanthanide Lewis acids and several chiral ligands revealed that $Y(OTf)_3$ and proline derived ligand **138** was optimal (data not shown). Upon further optimization it was discovered that achiral additives **139** and **212** increased ee's (Scheme 56, entries 2 and 3). Bulkier radicals were found to decrease the enantioselectivity (entries 4 and 5). Also, larger aryl substituents on the ligand gave similar ee's as observed for **138** (compare entries 1, 6, and 7).

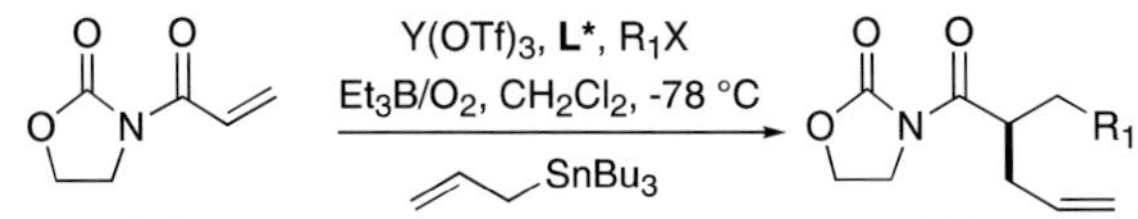

200 211

Entry	L*	Additive	R_1	Yield (%)	ee (%)
1	**138**	-	Et	61	70 (*S*)
2	**138**	**212**	Et	69	79 (*S*)
3	**138**	**139**	Et	71	80 (*S*)
4	**138**	**139**	*i*-Pr	67	68
5	**138**	**139**	*t*-Bu	46	42
6	**213**	-	Et	52	73 (*S*)
7	**213**	-	Et	60	73 (*S*)

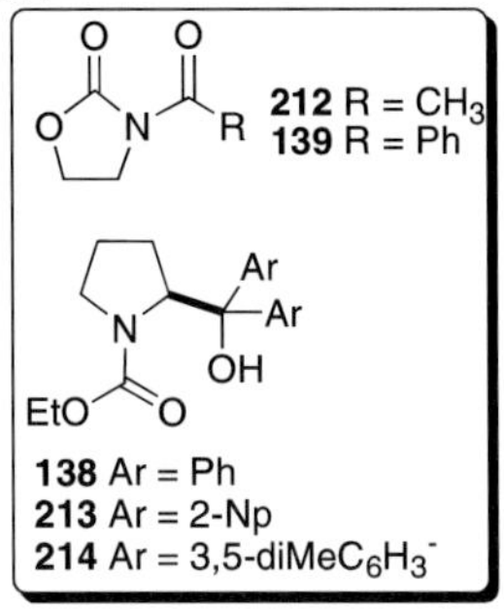

Scheme 56 Radical allylations using chiral lanthanide Lewis acids

6 Electron-Transfer Reactions

Reactions involving metals to generate radicals from oxygenated substrates such as aldehydes, ketones, and epoxides comprise a unique class in enantioselective radical reactions. Here, the radical generating reagent is bound to the radical precursor and remains bound to the reacting radical species allowing for stereocontrol in further reactions in the presence of a chiral ligand suitable for the metal. Some popular reagents in this class are titanocenes and samarium diiodide. Two recent reviews detail the development of transition metal reagents for catalysis in radical reactions [151, 152]. This section will summarize the developments in this field.

6.1 Ketyl Radical Reactions

The ability of HMPA in facilitating SmI_2 (Kagan reagent) mediated reactions has been well recognized and chiral ligands that have similar electron-donating capabilities have been tested in these reactions. Inanaga et al. have applied chiral samarium(II) complexes towards the hydrodimerization of acrylic acid amides **215** (Scheme 57) [153]. Dimerization of conjugated ketyl radicals in a ligand-controlled environment leads to the enantioselective formation of 3,4-*trans*-disubstituted adipamides **216**. Among the various bases that were evaluated, TMEDA proved to be the best. The reactions could be carried out with two equivalents of SmI_2 but as shown in the scheme, four equivalents were preferred due to the gradual decomposition of the chiral samarium complex even at $-78\,^{\circ}C$. The requirement of such excess of

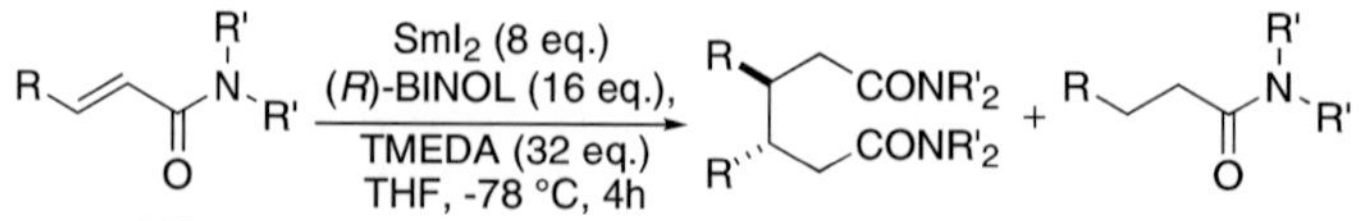

215 216 217

Entry	R	R'	Yield (%)	ee (%)
1	Me	Bn	70	71 (+)
2	Et	Bn	45	82 (+)
3	*n*-Pr	Bn	35	82 (+)
4	$BnCH_2$	Bn	20	85 (+)
5[a]	Me	Ph	55	44 (-)

[a] *dl:meso* =63:37

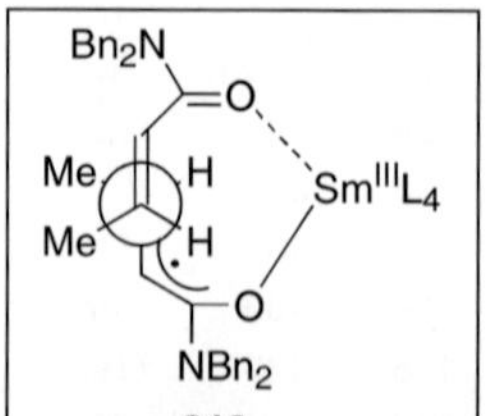

218

Scheme 57 Hydrodimerization through ketyl radicals

(*R*)-BINOL is discouraging but the ligand could be recovered in pure form following a simple work-up. The yield of **216** obtained in this reaction is rather low and is accompanied by a by-product **217**. No *meso* products were observed in the majority of cases. Good selectivities but moderate yields were obtained when *β*-substitution is with a linear alkyl group (entries 1–3). The efficiency of the reaction is lowered when bulky substituents are placed in the *β*-position and no homo-coupling was observed when *i*-propyl or *t*-butyl groups are present. Interestingly, when the amide substitution was changed from benzyl to phenyl, the *dl:meso* ratio was decreased and the opposite enantiomer was formed (entry 5). The higher reactivity of this substrate was postulated for this change in stereochemistry.

In cyclization reactions of ketyls with hydrazone, Skrydstrup and coworkers used different ligands to control the face selectivity in these coupling reactions [154]. Only low enantioselectivities (< 15%) and moderate yields (< 65%) were obtained for the *trans*-cyclized products (data not shown).

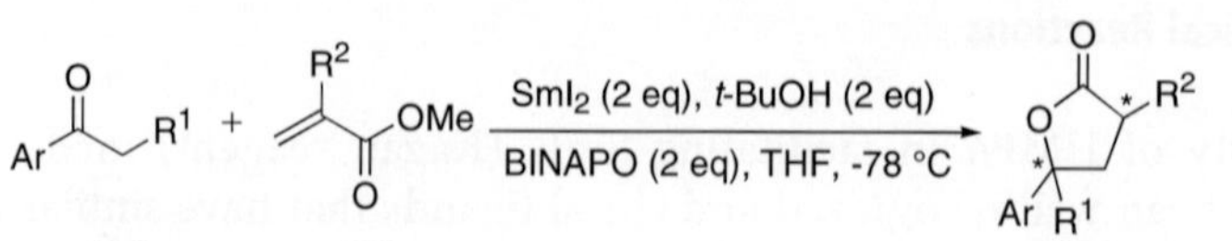

219 220 221

Ar	R^1	R^2	Yield (%)	*cis* : *trans*	ee (%) *cis/trans*
Ph	H	H	46		67
Ph	H	Me	42	66:34	89/55
Ph	Me	H	18		63

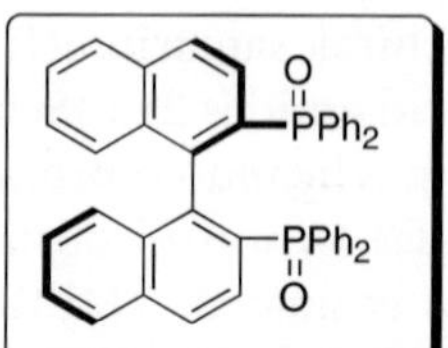

(*R*)- BINAPO, (*R*)-**222**

Scheme 58 Ketyl radical addition and cyclization

Mikami and co-workers demonstrated enantioselective addition of ketyl radicals generated using SmI_2 to olefins [155]. As shown in Scheme 58, the reductive coupling of acetophenone with methyl acrylate **220** in the presence of chiral 2,2′-bis(diphenylphosphinyl)-1,1′-binaphthyl (*R*-BINAPO) (*R*)-**222** gives somewhat low yields (mostly under 50%) but moderate to good levels of enantioselectivity (60–70% ee) for the γ-butyrolactone products **221**. Samarium diiodide is a one-electron donor and hence two equivalents of the metal are required in order for the reaction to proceed. The first electron donated from the samarium produces a chiral ketyl radical that undergoes enantioselective addition to the acrylate. The second electron donation then provides a samarium enolate intermediate that can potentially undergo stereoselective proton transfer in the formation of a second chiral center.

6.2 Pinacol Coupling

Reductive coupling of aldehydes using organometallic reagents to make pinacols is a powerful method. The most common metals for this process are titanium, vanadium, samarium, and niobium and of these titanium reagents are popular. The reaction involves generation of the ketyl radicals, which upon coupling provide the 1,2-diols. Various issues need to be considered regarding this reaction: control of both relative and absolute stereochemistry is required; in catalytic conditions, the product inhibition due to diol should be addressed; the reductant used for catalytic turnover further complicates the structure of the heterobimetallic reactive complex involved. Catalytic turnover can be achieved by cleaving the Ti – O bond using either TMSCl or proton. While TMSCl activates the aldehydes towards electron transfer, a general concern is the catalysis due to chlorosilane especially for less reactive substrates. Low valent titanium species can be either used directly (stoichiometric) or generated in situ (catalytic) using a reducing agent. Both these methods have been investigated in the enantioselective reactions (Scheme 59). Commercially available $TiCl_3$ has been used in stoichiometric conditions with (+)-dimethyltartarate **226** resulting in a drastic reduction in diastereoselectivity and very poor enantioselectivity (entry 2) [156]. $TiCl_2$ has been used along with diamine **227** providing diols with moderate ee's (entry 3). The added amines accelerate the reaction by making a homogeneous catalyst [157]. Experiments were conducted which suggested that there were two sets of particles with different sizes. This combined with the X-ray structure of the titanium-amine complex available in the literature led them to conclude that two species were present: one of them being a cluster of the $TiCl_2$-**227** and the other being a monomeric species with coordinated THF. It seems that the cluster leads to lower ee's: addition of tetrahydrothiophene resulted in higher selectivity (58% ee) [158]. Enders has used SMP, **228**, under similar condi-

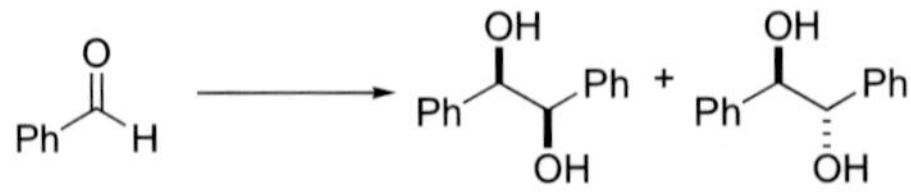

223 (*S*,*S*)-**224** **225**

Entry	Catalyst (eq.)	Conditions	Yield (%)	*dl:meso*	ee (%)	Ref.
1	$TiCl_3$ (1.0)	CH_2Cl_2/THF, rt, 30 min.	65	200:1	-	156
2	$TiCl_3$ (1.0) + **226** (0.5)	CH_2Cl_2/THF, rt, 30 min.	64	6.5:1	4	156
3	$TiCl_2$ (2.0) + **227** (4.0)	THF, rt, 8 h	88	14:1	40 (*S*,*S*)	157
4	$TiCl_2$ (2.0) + **228** (4.0)	THF, rt, 15 min.	31	4:1	65	159
5	**229** (0.1), Mn/TMSCl	THF, MS 4Å, rt, 24 h	-	7:1	60	160
6[a]	$TiCl_4(THF)_2$ (0.03), **230** (1.0), Mn (3.0)	CH_3CN, -10 °C	>95	49:1	91	162

[a] p-Methoxybenzaldehyde was used.

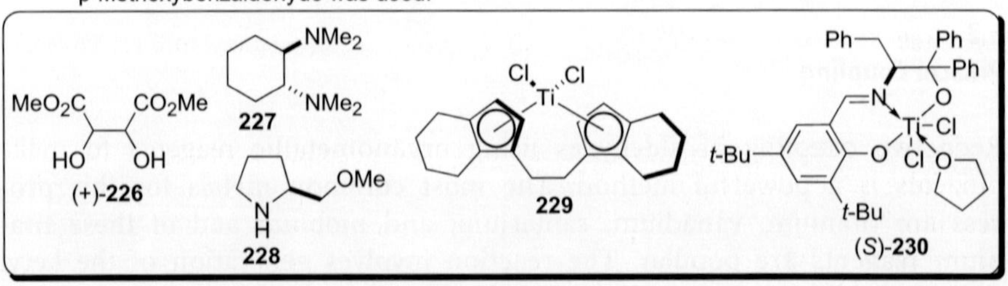

Scheme 59 Pinacol coupling using titanium catalysts

tions with low yields and moderate ee's (entry 4) [159]. Brintzinger's *ansa* metallocene **229** is the most successful chiral catalyst to date in pinacol coupling (entry 5) [160]. Other ligands have been screened including the commercially available (*R*,*R*)-(-)-*N*,*N*′-Bis(3,5-di-*tert*-butylsalicylidene)-1,2-cyclohexanediamine (salen) ligand [161]. A general observation is that the use of chelating ligands decreases the *dl:meso* selectivity in most cases. Recently, air-stable catalysts **230** derived from tridentate salen ligands and $TiCl_4$ were prepared and used in enantioselective pinacol coupling in both stoichiometric and catalytic amounts (entry 6) [162]. A survey of various reductants: Mn, Ce, SmI_2, and Zn indicated that manganese was the ideal choice at – 10 °C. Electron-deficient benzaldehydes were shown to give poor ee's for the pinacol products.

6.3 Epoxide Ring Opening

Titanium catalysts have long been used in electron transfer reactions involving epoxides, mostly as stoichiometric reagents. Gansäuer et al. have developed a catalytic version of these reactions using titanocenes along with zinc metal to generate the active catalyst (Scheme 60). In situ reduction of Ti(IV) with zinc metal provides Ti(III) species **231**, which coordinates

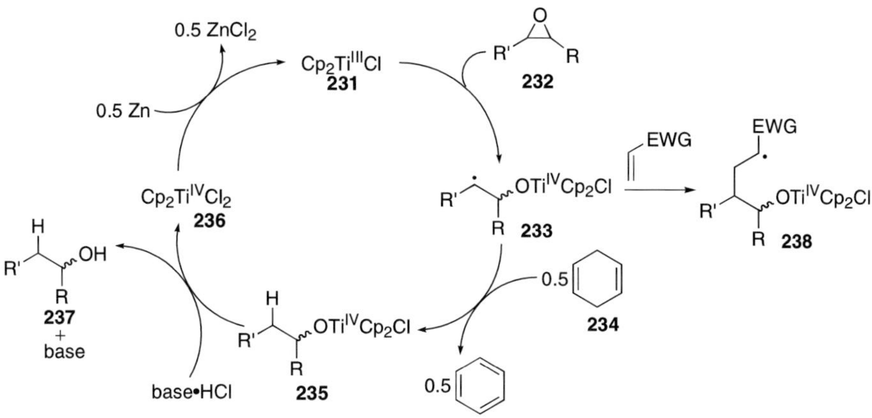

Scheme 60 Mechanism for titanocene-catalyzed epoxide ring opening and tandem reaction

to the epoxide **232**. The α-titanoxy radical **233** can be reduced with 1,4-cyclohexadiene generating **235** which is then protonated with collidine hydrochloride to provide the product and the Ti(IV) species which proceeds to the next catalytic cycle. On the other hand, **233** can add to an electron-deficient olefin to give **238**. Using chiral titanocenes allows for stereocontrol in both the ring opening as well as further reactions of the generated radical. Scheme 53 shows the results from the reductive opening of epoxides **239** to alcohols **240** [163, 164]. The catalysts derived from neo-menthol **241** and phenyl menthol **242** performed excellently providing products in > 93% ee (entries 3 and 4). Linear alkyl chains as substituents in the terminal ether of **239** are tolerated well whereas bulkier *t*-butyl groups decrease selectivity due to the steric disorientation of the discriminating groups (entries 3–6). The same catalysts were also utilized in the ring opening of cyclic epoxides **243** followed by addition to *t*-butyl acrylate **244** (Scheme 61). The *trans:cis* selectivities

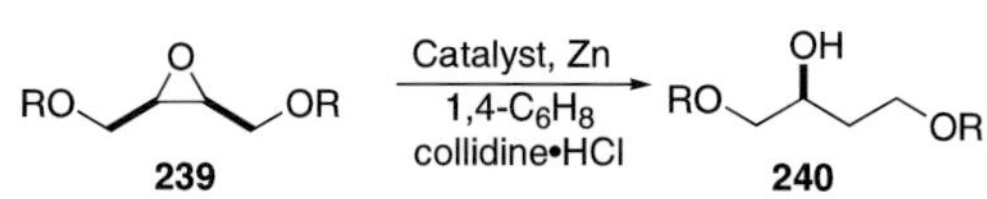

Entry	R	Catalyst	Yield (%)	ee (%)
1	Et	**241**	76	93
2	Et	**242**	72	94
3	*n*-Pr	**242**	60	92
4	*n*-Pr	**241**	70	91
5	*t*-Bu	**242**	68	74
6	*t*-Bu	**241**	66	86

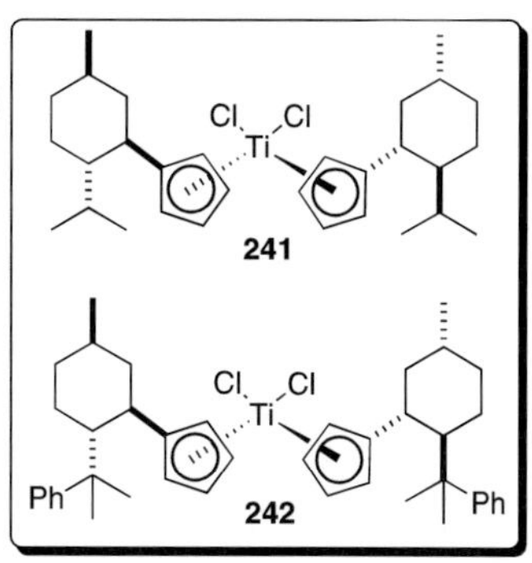

Scheme 61 Epoxide opening using titanocene catalysts

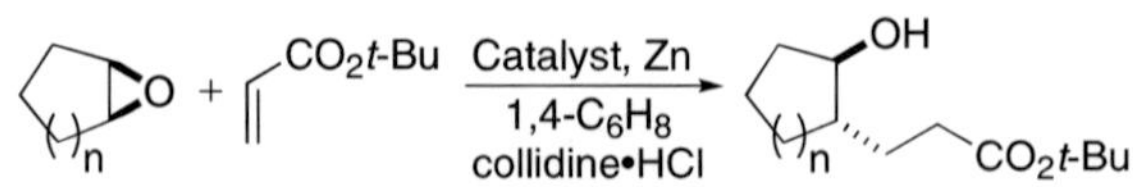

243 244 245

Entry	n	Catalyst	Yield (%)	ee (%)
1	1	**242**	69	74
2	1	**241**	72	81
3	2	**242**	61	82
4	3	**242**	65	82
5	3	**241**	78	80

Scheme 62 Tandem reactions with titanocene catalysts

were greater than 4 : 1 and good enantioselectivities for the *trans* isomers **245** were obtained. The selectivity was only slightly dependent on the ring size. Enantioselective cyclizations using this methodology would certainly lead to complex systems and will undoubtedly be useful in total synthesis.

6.4 Oxidative Coupling

The above three examples involved reactions where the electron transfer takes place from the metal to the organic substrate. The reverse scenario can also be used in radical reactions via oxidative generation of cationic radical species, which can undergo coupling reactions. Kurihara et al. have used chiral oxovanadium species as a one-electron transfer oxidant to silylenol ethers in a hetero-coupling process [165]. Treatment of **246** with a catalyst prepared in situ from $VOCl_3$/chiral alcohol/MS 4 Å followed by addition of **247** provided the coupling product **248** (Scheme 63). 8-Phenyl menthol **251** was found to be

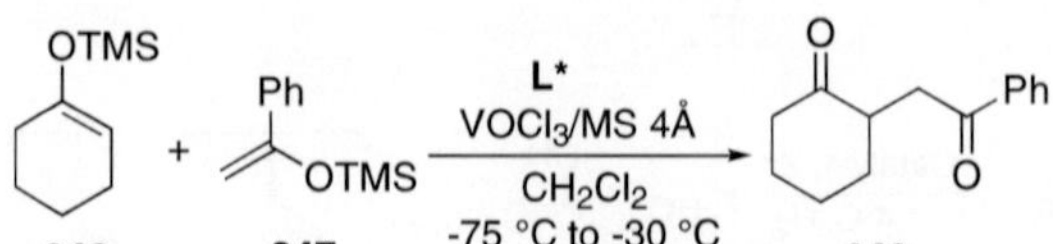

246 247 248

Entry	L*	Yield (%)	ee (%)
1	**249**	11	0
2	**250**	55	31
3	**251**	58	85

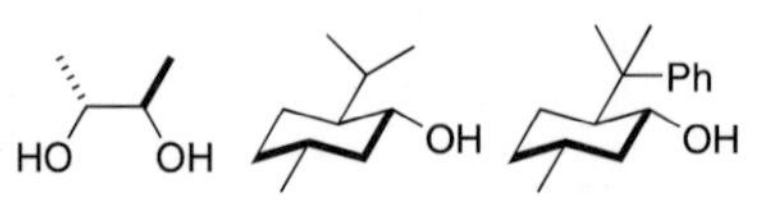

249 250 251

Scheme 63 Oxidative coupling of enol silanes

the best ligand for this process. MS 4 Å is essential to the success in generating the active catalyst. Although the results are good, a catalytic variant is not available at this time.

Oxidative homo-coupling of enolates from acyl oxazolidinones to give the corresponding dimers can be achieved in the presence of oxidants. Titanium and ytterbium enolates of **252** were coupled in the presence of a chiral diol or chiral bisoxazoline in the presence of ferrocenium cation **254** (Scheme 63) [166]. The amount of the meso dimer varied with the chiral ligand with a maximum of 5 : 1. TADDOL **172** performed best providing a 76% ee for the meso product. Ytterbium enolate gave a low ee of 34% with the same ligand.

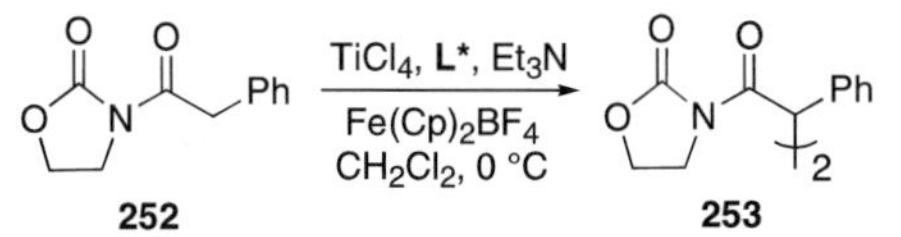

Entry	L*	Yield (%)	*dl:meso*	ee (%)
1	**255**	67	54:46	39 (*R*, *R*)
3	**183**	52	24:76	16 (*R*, *R*)
4	**172**	91	25:75	76 (*S*, *S*)

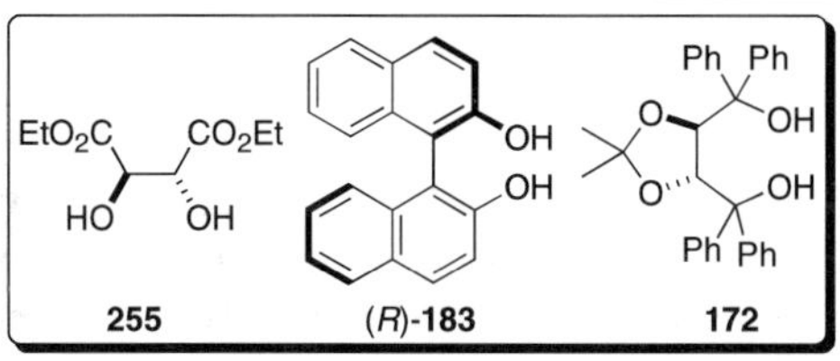

Scheme 64 Homo coupling of Ti/Yb enolates

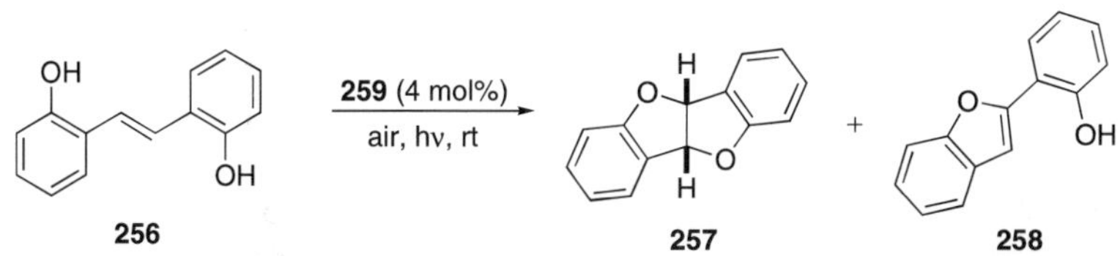

Entry	Solvent	Yield **257** (%)	ee (%)
1	chlorobenzene	79	81
2	*t*-BuOH	23	89
3	*i*-Pr_2O	12	85
4	toluene	78	82
5	tol:*t*-BuOH(2:3)	87	86

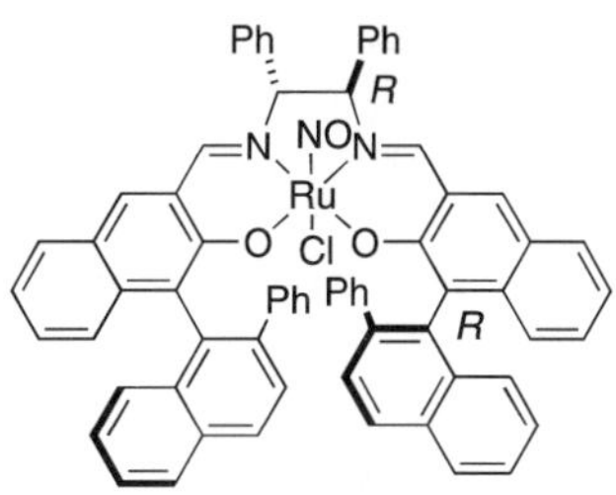

Scheme 65 Cyclizations under photolytic conditions

Cyclizations of dihydroxystilbene **256** using 4 mol % of chiral ruthenium complexes under photolytic conditions were investigated by Katsuki et al. (Scheme 65) [167]. Coordination of alcohols/phenols to Ru(IV) species generates a cation radical with concomitant reduction of metal to Ru(III). Cyclization of this oxygen radical followed by another cyclization provides the product **257**. Catalyst **259** provided 81% ee of the product in chlorobenzene solvent. Optimization of the solvent polarity led to a mixture of toluene and *t*-butanol in 2 : 3 ratio as the ideal solvent. Substituents on the phenyl rings led to a decrease in selectivity. Low yields were due to the by-product **258**.

7 Conclusion

The development of enantioselective radical reactions outlined in this review is remarkable. Most of the processes described in this review involve formation of either C – C or C – H bonds (except for halogen atom transfer and one example of the C – O bond: Schenk-ene reaction). There are examples of halogen atom transfer but no examples of enantioselective group transfer were found in the literature. Additionally, enantioselective cyclizations have been investigated to a limited extent. Establishing multiple stereocenters in a single reaction is intriguing and until now, the record stands at four stereocenters for two bond construction. There is need to further improve applications of radical processes by introducing more functional groups and to increase the complexity of the substrates used or the radicals that are being added. Radical reactions in alternate media, the use of polymer-supported chiral reagents, and enantioselective transformations on solid-support are awaiting exploration.

References

1. Renaud P, Sibi MP (eds) (2001) Radicals in Organic Synthesis, vols 1 and 2. Wiley, Weinheim
2. Parsons AF (2000) An Introduction to Free Radical Chemistry. Blackwell Science, Oxford
3. Alfassi ZB (1999) General Aspects of the Chemistry of Radicals. Wiley, New York
4. Fossey J, Lefort D, Sorba J (1995) Free Radical in Organic Chemistry. Wiley, New York
5. Giese B (1986) Radicals in Organic Synthesis. Formation of Carbon–Carbon Bonds. Pergamon, Oxford
6. Curran DP (1991) Radical Addition Reactions. In: Trost BM, Fleming I, Semmelheck MF (eds) Comprehensive Organic Synthesis. Pergamon, Oxford, p 715–777
7. Curran DP, Porter NA, Giese B (1995) Stereochemistry of Radical Reactions. Wiley, Weinheim
8. Smadja W (1994) Synlett 1–26

9. Beckwith AL (1993) Chem Soc Rev 22:143–151
10. Porter NA, Giese B, Curran DP (1991) Acc Chem Res 24:296–301
11. Giese B (1989) Angew Chem Int Ed Engl 28:969–980
12. Sibi MP, Manyem S, Zimmerman J (2003) Chem Rev 103:3263–3295
13. Sibi MP, Rheault TR (2001) In: Renaud P, Sibi MP (eds) Radicals in Organic Synthesis. Wiley, Weinheim, p 461–478
14. Parsons AF, Bar G (2003) Chem Soc Rev 251–263
15. Sibi MP, Porter NA (1999) Acc Chem Res 32:163–171
16. Gawley RE, Aube J (1996) Principles of Asymmetric Synthesis. Pergamon, Oxford, p 14
17. Perlmutter P (1992) Conjugate Addition Reactions in Organic Synthesis. Pergamon Press, Oxford
18. Sibi MP, Manyem S (2000) Tetrahedron 56:8033–8061
19. Jasperse CP, Curran DP, Fevig TL (1991) Chem Rev 91:1237–1286
20. Curran DP (2000) Aldrichimica Acta 33:104–110
21. Murakata M, Tsutsui H, Takeuchi N, Hoshino O (1999) Tetrahedron 55:10295–10304
22. Urabe H, Yamashita K, Suzuki K, Kobayashi K, Sato F (1995) J Org Chem 60:3576–3577
23. Sibi MP, Sausker JB (2002) J Am Chem Soc 124:984–991
24. Sugimoto H, Nakamura S, Watanabe Y, Toru T (2003) Tetrahedron: Asymmetry 14:3043–3055
25. Sibi MP, Asano Y, Sausker JB (2001) Angew Chem Int Ed 40:1293–1296
26. Liu M, Sibi MP (2002) Tetrahedron 58:7991–8035
27. Juaristi E, Lopez-Ruiz H (1999) Curr Med Chem 6:983–1004
28. Davies HML, Venkataramani C (2002) Angew Chem Int Ed 41:2197–2199
29. Duursma A, Minnaard AJ, Feringa BL (2003) J Am Chem Soc 125:3700–3701
30. Sibi MP, Patil K (2004) Angew Chem Int Ed 43:1235–1238
31. Basavaiah D, Rao AJ, Satyanarayana T (2003) Chem Rev 103:811–891
32. Langer P (2003) Org Synth Highlights V, p 165–177
33. Langer P (2000) Angew Chem Int Ed 39:3049–3052
34. Ciganek E (1997) Organic Reactions. New York 51:201–350
35. Basavaiah D, Rao PD, Hyma RS (1996) Tetrahedron 52:8001–8062
36. Sibi MP, Zimmerman J, Rheault T (2003) Angew Chem Int Ed 43:4521–4523
37. Garner P, Anderson JT, Cox PB, Klippenstein SJ, Leslie R, Scardovi N (2002) J Org Chem 67:6195–6209
38. Lee E, Tae JS, Lee C, Park CM (1993) Tetrahedron Lett 34:4831–4834
39. Sibi MP, Patil K (2005) Org Lett 7:1453–1456
40. Aechtner T, Dressel M, Bach T (2004) Angew Chem Int Ed 43:5849–5851
41. Buckmelter AJ, Kim AI, Rychnovsky SD (2000) J Am Chem Soc 122:9386–9390
42. Dalgard JE, Rychnovsky SD (2004) Org Lett 6:2713–2716
43. Ohno A, Ikeguchi M, Kimura T, Oka S (1979) J Am Chem Soc 101:7036–7040
44. Tanner DD, Kharrat A (1988) J Am Chem Soc 110:2968–2970
45. Gielen M, Mokhtar-Jamai H (1975) Bull Chem Soc Belg 84:197–202
46. Gielen M, Tondeur Y (1979) J Organomet Chem 169:265–281
47. Schumann H, Pachaly B, Schütze BC (1984) J Organomet Chem 169:145–152
48. Nanni D, Curran DP (1996) Tetrahedron: Asymmetry 7:2417–2422
49. Blumenstein M, Schwarzkopf K, Metzger JO (1997) Angew Chem Int Ed Engl 36:235–236
50. Blumenstein M, Lemmler M, Hayen A, Metzger JO (2003) Tetrahedron: Asymmetry 14:3069–3077
51. Schwarzkopf K, Blumenstein M, Hayen A, Metzger JO (1998) Eur J Org Chem, p 177–181

52. Helliwell M, Thomas EJ, Townsend LA (2002) J Chem Soc Perkin Trans 1:1286–1296
53. Dakternieks D, Dunn K, Perchyonok VT, Schiesser CH (1999) Chem Commun, p 1665–1666
54. Dakternieks D, Perchyonok VT, Schiesser CH (2003) Tetrahedron: Asymmetry 14:3057–3068
55. Dakternieks D, Dunn K, Henry DJ, Schiesser CH, Tiekink ERT (1999) Organomet 18:3342–3347
56. Dakternieks D, Dunn K, Schiesser CH, Tiekink ERT (2000) J Organomet Chem 605:209–220
57. Zeng L, Perchyonok VT, Schiesser CH (2004) Tetrahedron: Asymmetry 15:995–999
58. Dakternieks D, Schiesser CH (2001) Aust J Chem 54:89–91
59. Schiesser CH, Skidmore MA, White JM (2001) Aust J Chem 54:199–204
60. Kang J, Kim TH (2003) Bull Korean Chem Soc 24:1055–1056
61. Zeng L, Dakternieks D, Duthie A, Perchyonok VT, Schiesser CH (2004) Tetrahedron: Asymmetry 15:2547–2554
62. Curran DP, Gualtieri G (2001) Synlett 1038–1041
63. Gualtieri G, Geib SJ, Curran DP (2003) J Org Chem 68:5013–5019
64. Mok PLH, Roberts BP (1992) Tetrahedron Lett 33:7249–7252
65. Mok PLH, Roberts BP, McKetty PT (1993) J Chem Soc Perkin Trans 2:665–674
66. Dang H-S, Diart V, Roberts BP (1994) J Chem Soc Perkin Trans 1:1033–1041
67. Dang H-S, Diart V, Roberts BP, Tocher DA (1994) J Chem Soc Perkin Trans 2:1039–1045
68. Haque MB, Roberts BP (1996) Tetrahedron Lett 37:9123–9126
69. Haque MB, Roberts BP, Tocher DA (1998) J Chem Soc Perkin Trans 1:2881–2889
70. Cai Y, Roberts BP, Tocher DA (2002) J Chem Soc Perkin Trans 1:1376–1386
71. Dang H-S, Roberts BP (1995) Tetrahedron Lett 36:3731–3734
72. Dang H-S, Kim K-M, Roberts BP (1998) Chem Commun 1413–1414
73. Byers J (2001) In: Renaud P, Sibi MP (eds) Radicals in Organic Synthesis, vol 1. Wiley, Weinheim, p 72–89
74. Kameyama M, Kamigata N (1989) Bull Chem Soc Jpn 62:648–650
75. Kameyama M, Kamigata N, Kobayashi M (1987) J Org Chem 52:3312–3316
76. Kameyama M, Kamigata N (1987) Bull Chem Soc Jpn 60:3687–3691
77. Murai S, Sugise R, Sonoda N (1981) Angew Chem Int Ed Engl 20:475–476
78. Mero CL, Porter NA (1999) J Am Chem Soc 121:5155–5160
79. Brunner H, Bluchel C, Doyle MP (1997) J Organomet Chem 541:89–95
80. Yang D, Gu S, Yan YL, Zhu NY, Cheung KK (2001) J Am Chem Soc 123:8612–8613
81. Yang D, Gu S, Yan YL, Zhao HW, Zhu NY (2002) Angew Chem Int Ed 41:3014–3017
82. Miyabe H, Ushiro C, Ueda M, Yamakawa K, Naito T (2000) J Org Chem 65:176–185
83. Halland N, Jørgensen KA (2001) J Chem Soc Perkin Trans 1:1290–1295
84. Friestad GK, Shen Y, Ruggles EL (2003) Angew Chem Int Ed 42:5061–5063
85. Sibi MP, Manyem S (2000) Tetrahedron 56:8033–8061
86. Sibi MP, Ji JG, Wu JH, Gürtler S, Porter NA (1996) J Am Chem Soc 118:9200–9201
87. Sibi MP, Ji JG (1997) J Org Chem 62:3800–3801
88. Sibi MP, Petrovic G (2003) Tetrahedron: Asymmetry 14:2879–2882
89. Iserloh U, Curran DP, Kanemasa S (1999) Tetrahedron: Asymmetry 10:2417–2428
90. Murakata M, Tsutsui H, Hoshino O (2001) Org Lett 3:299–302
91. Sibi MP, Shay JJ, Ji JG (1997) Tetrahedron Lett 38:5955–5958
92. Sibi MP, Prabagaran N (2004) Synlett 2421–2424
93. Sibi MP, Manyem S (2002) Org Lett 4:2929–2932
94. Kobayashi S, Sugiura M, Kitagawa H, Lam WWL (2002) Chem Rev 102:2227–2302

95. Sibi MP, He L (2004) Org Lett 6:1749–1752
96. Sibi MP, Guerrero M (2005) Synthesis 1529–1532
97. Otera J (ed) (2000) Modern Carbonyl Chemistry. Wiley, Weinheim
98. Alcaide B, Almendros P (2002) Eur J Org Chem 65:1595–1601
99. Johnson JS, Evans DA (2000) Acc Chem Res 33:325–335
100. Trost BM, Ito H (2000) J Am Chem Soc 122:12003–12004
101. List B, Lerner RA, Barbas CF (2000) J Am Chem Soc 122:2395–2396
102. Singer RA, Carreira EM (1995) J Am Chem Soc 117:12360–12361
103. Garner P, Anderson JT, Cox PB, Klippenstein SJ, Leslie R, Scardovi N (2002) J Org Chem 67:6195–6209
104. Garner P, Anderson JT (1999) Org Lett 1:1057–1059
105. Guindon Y, Denis RC (1998) Tetrahedron Lett 39:339–342
106. Lee E, Tae JS, Lee C, Park CM (1993) Tetrahedron Lett 34:4831–4834
107. Lee E, Yoo S-K, Choo H, Song HY (1998) Tetrahedron Lett 39:317–318
108. Evans PA, Murthy VS, Roseman JD, Rheingold AL (1999) Angew Chem Int Ed 38:3175–3177
109. Lee E, Choi SJ (1999) Org Lett 1:1127–1128
110. Sibi MP, Liu M (2001) Curr Org Chem 5:719–755
111. Yamada K-I, Arai T, Sasai H, Shibasaki M (1998) J Org Chem 63:3666–3672
112. Arai T, Sasai H, Aoe K, Okamura K, Date T, Shibasaki M (1996) Angew Chem Int Ed Engl 35:104–106
113. Naasz R, Arnold LA, Pineschi M, Keller E, Feringa BL (1999) J Am Chem Soc 121:1104–1105
114. Alexakis A, Trevitt GP, Bernardinelli G (2001) J Am Chem Soc 123:4358–4359
115. Arnold LA, Naasz R, Minnaard AJ, Feringa BL (2001) J Am Chem Soc 123:5841–5842
116. Doi H, Sakai T, Iguchi M, Yamada K-I, Tomioka K (2003) J Am Chem Soc 125:2886–2887
117. Nishimura K, Tomioka K (2002) J Org Chem 67:431–434
118. Curran DP, Porter NA, Giese B (1995) Stereochemistry of Radical Reactions. Wiley, Weinheim
119. Durkin K, Liotta D, Rancourt J, Lavallée J-F, Boisvert L, Guindon Y (1992) J Am Chem Soc 114:4912–4914
120. Guindon Y, Houde K, Prevost M, Cardinal-David B, Landry SR, Daoust B, Bencheqroun M, Guerin B (2001) J Am Chem Soc 123:8496–8501
121. Curran DP, Abraham AC (1993) Tetrahedron 49:4821–4840
122. Curran DP, Ramamoorthy PS (1993) Tetrahedron 49:4841–4858
123. Curran DP, Geib S, De Mello N (1999) Tetrahedron 55:5681–5704
124. Kopping B, Chatgilialoglu C, Zehnder M, Giese B (1992) J Org Chem 57:3994–4000
125. Sibi MP, Petrovic G, Zimmerman J (2005) J Am Chem Soc 127:2390–2391
126. Evans DA, Tedrow JS, Shaw JT, Downey CW (2002) J Am Chem Soc 124:392–393
127. Denmark SE, Wynn T, Beutner GL (2002) J Am Chem Soc 124:13405–13407
128. Ghosh AK, Kim J-H (2003) Org Lett 5:1063–1066
129. Kiyooka S (2003) Tetrahedron: Asymmetry 14:2897–2910
130. Walker MA, Heathcock CH (1991) J Org Chem 56:5747–5750
131. Corey EJ, Kim SS (1990) J Am Chem Soc 112:4976–4977
132. Evans DA, Downey CW, Shaw JT, Tedrow JS (2002) Org Lett 4:1127–1130
133. He L, Srikanth GSC, Castle SL (2005) J Org Chem 70:8140–8147
134. Nishida M, Hayashi H, Nishida A, Kawahara N (1996) Chem Commun 579–580
135. Hiroi K, Ishii M (2000) Tetrahedron Lett 41:7071–7074
136. Curran DP, Liu W, Chen CH-T (1999) J Am Chem Soc 121:11012–11013

137. Petit M, Lapierre AJB, Curran DP (2005) J Am Chem Soc 127, ASAP
138. Rosenstein IJ (2001) In: Renaud P, Sibi MP (eds) Radicals in Organic Synthesis, vol 1. Wiley, Weinheim, p 50–71
139. Keck GE, Yates JB (1982) J Org Chem 47:3590–3591
140. Porter NA, Wu JHL, Zhang GR, Reed AD (1997) J Org Chem 62:6702–6703
141. Fhal AR, Renaud P (1997) Tetrahedron Lett 38:2661–2664
142. Porter NA, Feng H, Kavrakova IK (1999) Tetrahedron Lett 40:6713–6716
143. Murakata M, Jono T, Mizuno Y, Hoshino O (1997) J Am Chem Soc 119:11713–11714
144. Murakata M, Jono T, Hoshino O (1998) Tetrahedron: Asymmetry 9:2087–2092
145. Wu JH, Radinov R, Porter NA (1995) J Am Chem Soc 117:11029–11030
146. Wu JH, Zhang GR, Porter NA (1997) Tetrahedron Lett 38:2067–2070
147. Sibi MP, Ji JG (1996) J Am Chem Soc 118:3063
148. Watanabe Y, Mase N, Furue R, Toru T (2001) Tetrahedron Lett 42:2981–2984
149. Sibi MP, Chen J (2001) J Am Chem Soc 123:9472–9473
150. Sibi MP, Manyem S, Subramaniam R (2003) Tetrahedron 59:10575–10580
151. Gansäuer A, Bluhm H (2000) Chem Rev 100:2771–2788
152. Gansäuer A, Narayan S (2002) Adv Synth Catal 344:465–475
153. Kikukawa T, Hanamoto T, Inanaga J (1999) Tetrahedron Lett 40:7497–7500
154. Riber D, Hazell R, Skrydstrup T (2000) J Org Chem 65:5382–5390
155. Mikami K, Yamaoka M (1998) Tetrahedron Lett 39:4501–4504
156. Clerici A, Clerici L, Porta O (1996) Tetrahedron Lett 37:3035–3038
157. Matsubara S, Hashimoto Y, Okano T, Utimoto K (1999) Synlett 1411
158. Hashimoto Y, Mizuno U, Matsuoka H, Miyahara T, Takakura M, Yoshimoto M, Oshima K, Utimoto K, Matsubara S (2001) J Am Chem Soc 123:1503–1504
159. Enders D, Ullrich EC (2000) Tetrahedron: Asymmetry 11:3861–3865
160. Dunlap MS, Nicholas KM (1999) Synth Commun 27:1097–1106
161. Bandini M, Cozzi PG, Morganti S, Umani-Ronchi A (1999) Tetrahedron Lett 40:1997–2000
162. Bensari A, Renaud JL, Riant O (2001) Org Lett 3:3863–3865
163. Gansäuer A, Bluhm H, Lauterbach T (2001) Adv Synth Catal 343:785–787
164. Gansäuer A, Lauterbach T, Bluhm H, Noltemeyer M (1999) Angew Chem Int Ed 38:2909–2910
165. Kurihara M, Hayashi T, Miyata N (2001) Chem Lett 1324–1325
166. Nguyen PQ, Schäfer HJ (2001) Org Lett 3:2993–2995
167. Masutani K, Irie R, Katsuki T (2002) Chem Lett 36–37

Top Curr Chem (2006) 263: 163–189
DOI 10.1007/128_028

Published online: 23 February 2006

Radical Stability—A Theoretical Perspective

H. Zipse

Department of Chemistry and Biochemistry, LMU München, Butenandtstrasse 5–13, 81377 München, Germany
zipse@cup.uni-muenchen.de

Abstract The thermodynamic stability of radicals as defined through isodesmic hydrogen transfer reactions has been explored at a variety of theoretical levels. Radical stabilization energies (RSEs) derived from single point calculations at the ROMP2/6-311+G(3df,2p)//UBecke3LYP/6-31G(d) level of theory in combination with scaled zero point vibrational energies calculated at the UBecke3LYP/6-31G(d) level have been determined for a broad variety of systems. For the three radical types considered in this study (carbon-, nitrogen-, and oxygen-centered radicals) the radical stabilization energy (RSE) depends on the same fundamental effects such as resonance stabilization of the unpaired spin, electron donation through adjacent alkyl groups or lone pairs, and through inductive electron donation/electron withdrawal. The influence of ring strain effects as well as the synergistic combination of individual substituent effects have also been explored.

Keywords DFT calculations · Isodesmic equations · Radicals · ROMP2 calculations · Thermodynamic stability

Abbreviations

BDE	bond dissociation energy
CBS	complete basis set theory
DFT	density functional theory
G3	Gaussian-3 theory
HLC	higher level correction terms
ROMP2	2nd order restricted open shell Møller–Plesset theory
RSE	radical stabilization energy
W1	Weizmann-1 theory

W2 Weizmann-2 theory
ZPVE zero point vibrational energy

1 Introduction

Together with the detailed knowledge of rate constants for individual elementary steps the thermodynamic stability of radicals provides a quantitative basis for the rational design of radical reactions for organic synthesis. In principle the stability of radicals can be defined in kinetic as well as thermodynamic terms [1]. The theoretical prediction of thermodynamic stabilities is particularly attractive because it allows for the direct comparison of radicals of widely different structure and electronic characteristics. The definition of thermodynamic stability is in general bound to an arbitrarily chosen reference system. While this is also true for the definition of radical stability, the use of the smallest organic radical, the methyl radical (**1**), and its closed shell parent compound, methane (**2**), as the reference systems appears to be the most meaningful (and also the most widespread) choice [2–5]. With this reference the determination of radical stability equates to the prediction of the reaction enthalpy for reaction Eq. 1, most commonly termed the "radical stabilization energy" (RSE):

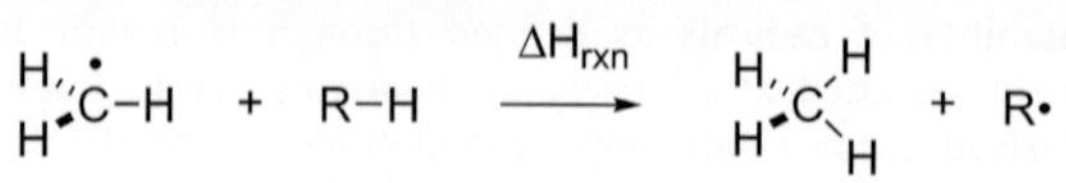

1 **2** (1)

The reaction enthalpy and thus the RSE will be negative for all radicals, which are more stable than the methyl radical. Equation 1 describes nothing else but the difference in the bond dissociation energies (BDE) of $CH_3 - H$ and $R - H$, but avoids most of the technical complications involved in the determination of absolute BDEs. It can thus be expected that even moderately accurate theoretical methods give reasonable RSE values, while this is not so for the prediction of absolute BDEs. In principle, the isodesmic reaction described in Eq. 1 lends itself to all types of carbon-centered radicals. However, the error compensation responsible for the success of isodesmic equations becomes less effective with increasingly different electronic characteristics of the $C - H$ bond in methane and the $R - H$ bond. As a consequence the stability of σ-radicals located at sp^2 hybridized carbon atoms may best be described relative to the vinyl radical **3** and ethylene **4**:

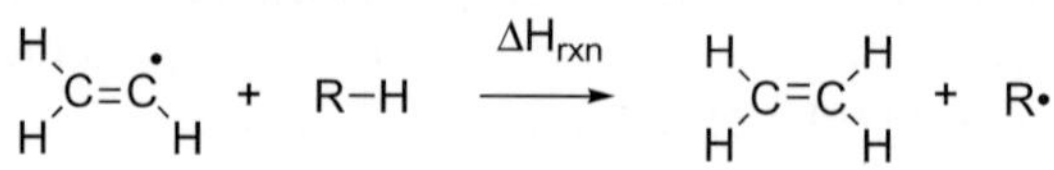

3 **4** (2)

The stability of heteroatom-centered radicals can be defined relative to reference systems sharing the same type of radical center. The stability of nitrogen-centered radicals may, for example, be defined relative to ammonia (**6**) and the amino radical (**5**), and the most obvious choice for oxygen-centered radicals are the hydroxyl radical (**7**) and water (**8**):

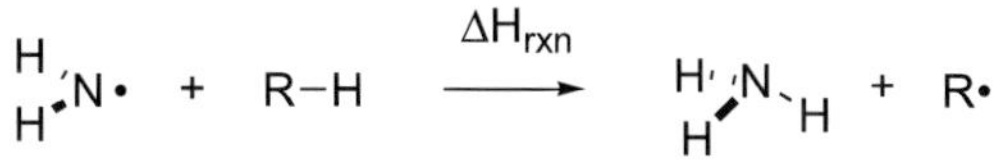

5 **6** (3)

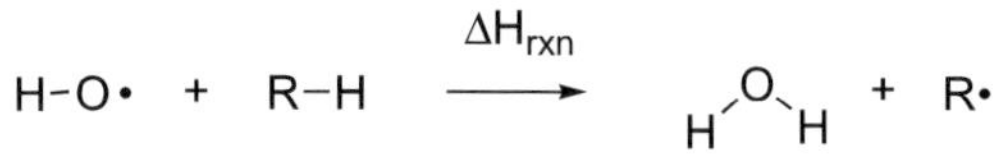

7 **8** (4)

The stabilities of radicals can, of course, also be defined through non-isodesmic reactions [6]. The benefits of isodesmic error compensation are lost in these cases and a higher level of theory may be required for an accurate prediction of stabilization energies. Also, the reaction enthalpies of hydrogen-transfer reactions are by no means the only way to define the thermodynamic stability of radicals (see example in Sect. 1.3) and other reference systems (e.g. those based on bond energy terms) are certainly also possible [9]. One concern with Eq. 1 as the defining equation of radical stability has been that it reflects substituent effects on both the radical and its closed shell parent. In a number of studies the substituent effects on closed shell compounds have therefore been estimated separately (e.g. through appropriate isodesmic reactions) and then subtracted from the RSE values calculated according to Eq. 1 in order to arrive at the true substituent effects on the radical. While these corrected RSE values are certainly more appropriate for a discussion of substituent effect on radicals (see Sect. 1.3 for some examples), they cease to reflect the X – H bond energy difference between two systems [8].

1.1 Theoretical Methods

Early determinations of RSE values employed unrestricted Hartree-Fock (UHF) theory in combination with 3-21G [9] or 4-31G [10] basis sets to evaluate the RSE according to Eq. 1. The appropriate consideration of correlation effects, the avoidance of spin contamination, and the treatment of thermochemical corrections have in detail been studied in the following, in particular by Bauschlicher [11], Coote [12–14], Morokuma [15–18], and Radom [19–25]. Highly accurate RSE and BDE results can be obtained with high level compound methods such as the G2 [26–30] and G3 [31–34] schemes (and variants thereof [11, 15–18]), as well as extrapolation methods such as the CBS schemes [35, 36], W1, or W2 [37–39]. Generally, the accurate characterization of oxygen- and nitrogen-centered radicals is somewhat more demanding than calculations on carbon-centered radicals [40–42].

Moderately accurate RSE values can be obtained for a much larger selection of radicals using DFT-optimized geometries (e.g. at the Becke3LYP/6-31G(d) level of theory) in combination with single point calculations using larger basis sets at either DFT [43–46] or MP2 level. Very promising results have recently been reported for carbon-centered radicals using ROMP2/6-311+G(3df,2p)//UBecke3LYP/6-31G(d) single point calculations in combination with zero point vibrational energies (ZPVE) calculated at the UBecke3LYP/6-31G(d) level and scaled by 0.9806 [12, 25]. This model will in the following be termed "ROMP2". A model of slightly better accuracy termed "G3(MP2)-RAD" [22] has been tailored by Radom and coworkers for applications in radical chemistry and is based on the G3(MP2) scheme by Curtiss and coworkers [31–33]. It involves the same geometries and ZPVE as the "ROMP2" model described above, but determines electronic energies through a series of single point calculations at URCCSD(T)/6-31G(d), ROMP2/G3MP2large, and ROMP2/6-31G(d) level in combination with higher level correction (HLC) terms derived empirically. It must be emphasized that both models generate RSE values at a temperature of 0 K, but account for differences in ZPVE. Temperature corrections of RSE values up to 298 K are only moderately relevant and will also be neglected here.

RSE values can also be calculated from experimentally measured X – H bond dissociation energies or heats of formation (where available). In order to be directly comparable to the RSE values calculated at the "ROMP2" or "G3(MP2)-RAD" level described above, this requires thermochemical data for the species in Eqs. 1–4 at 0 K. One straightforward approach is the back correction of experimentally measured heats of formation at 298.15 K to 0 K values using thermochemical corrections calculated using the rigid rotor/harmonic oscillator model in combination with scaled DFT or UMP2 frequencies [19, 23].

1.2 Alkyl Radicals with One Substituent

Numerous reports published in recent years have focused on carbon-centered radicals derived from compounds with selected substitution patterns such as alkanes [40, 43, 47], halogenated alkanes [43, 48, 49, 51–57], alkenes [19], benzene derivatives [43, 47], ethers [51, 58], aldehydes [48], amines [10, 59], amino acids [23, 60–67] etc. Particularly significant advances have been made in the theoretical treatment of radicals occurring in polymer chemistry and biological chemistry. The stabilization of radicals in all of these compounds is due to the interaction of the molecular orbital carrying the unpaired electron with energetically and spatially adjacent molecular orbitals, and four typical scenarios appear to cover all known cases [20].

The most common and also most effective mechanism of radical stabilization involves the resonant delocalization of the unpaired spin into an adjacent π system, the allyl radical being the prototype case. A minimal orbital interaction diagram describing this type of stabilization mechanism involves the unpaired electron located in a π-type orbital at the formal radical center and the π- and π^*-orbitals of the π system (Scheme 1).

This type of interaction involves three electrons and three centers and is strongly stabilizing. Taking the allyl radical as an example the stabilization amounts to -77.5 kJ/mol at the ROMP2 level (Table 1). This type of stabilization mechanism is present in practically all radical substituents containing a π system. The stabilization through heteroatom-substituted π systems such as $-CN$ (-31.1 kJ/mol), $-CHO$ (-32.2 kJ/mol), $-COCH_3$ (-28.2 kJ/mol), $-CO-C_6H_5$ (-26.5 kJ/mol), $-CO-NH_2$ (-20.3 kJ/mol), $-COOH$ (-20.0 kJ/mol), $-NO_2$ (-11.7 kJ/mol) is less effective as compared to the all-carbon analogs and the stabilization decreases steadily with increasingly electronegative components of the substituent. With respect to the orbital interaction diagram shown in Scheme 1 this can best be understood as the consequence of the lowering of π and π^* orbitals with increasingly electronegative heteroatom substitution. The stabilization energies are also lower for other all-carbon π systems such as the phenyl group (-50.4 kJ/mol) or the C–C triple bond in the propargyl radical (-50.6 kJ/mol). The radical structure reflects this stabilizing interaction through formation of a typically fully planar radical center as well as a rather short bond between the radical center and stabilizing substituent (as compared to the corresponding closed shell compound). The allyl radical as the most typical example exhibits two C–C bonds of 138.6 pm length, while the C–C single bond in propene amounts to 150.2 pm (both at the B3LYP/6-31G(d) level of theory).

The stabilization of carbon-centered radicals through alkyl groups is due to a closely similar orbital interaction as that shown for π systems (Scheme 2).

The π orbitals constructed through combination of individual C–H bonds can interact with the unpaired spin in much the same way as seen before for

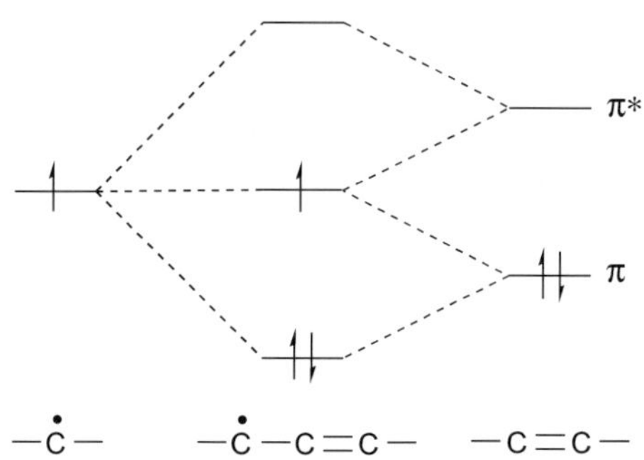

Scheme 1

Table 1 Radical stabilization energies (in kJ/mol) of monosubstituted methyl radicals at 0 K according to Eq. 1

Radical	RSE (ROMP2)	RSE (other)	Refs.
$^{\cdot}CH_2-NH_3^+$	+ 18.3	—	[10, 93, 94]
$^{\cdot}CH_2-SH_2^+$	+ 12.8	—	[10, 93, 94]
$^{\cdot}CH_2-CF_3$	+ 8.1	+ 6.7 (CBS-4) + 7.7 (G3(MP2)-RAD)	[7, 10, 20, 48, 51, 57]
$^{\cdot}CH_2-PH_3^+$	+ 7.0	—	[10, 93, 94]
$^{\cdot}CH_2-SO_2-CH_3$	+ 6.3	—	[10, 95, 96]
$^{\cdot}CH_2-CF_2-CF_3$	+ 5.3	+ 5.4 (CBS-4) + 4.9 (G3(MP2)-RAD)	[20, 57]
$^{\cdot}CH_2-CF_2-H$	+ 3.0	—	[10, 51]
$^{\cdot}CH_2-CCl_3$	+ 2.6	—	
$^{\cdot}CH_3$	0.0	0.0	[2, 4, 5, 7, 12, 18, 21–24, 31, 43, 44, 56, 57, 60, 72, 74, 81, 88, 93, 95, 97–99]
$^{\cdot}CH_2-CCl_2H$	- 1.8	—	
$^{\cdot}CH_2-SO-CH_3$	- 4.1	—	[10]
$^{\cdot}CH_2-CH_2-F$	- 5.5	—	[4, 10, 51, 134]
$^{\cdot}CH_2-C(CH_3)_3$	- 6.5	—	[100, 134]
$^{\cdot}CH_2-CH_2-Cl$	- 6.7	—	[4]
$^{\cdot}CH_2-CH_2-OH$	- 8.5	—	[50, 135]
$^{\cdot}CH_2-CH_2-C_6H_5$	- 9.2	- 10.6 (G3(MP2)-RAD)	[13, 93, 97]
$^{\cdot}CH_2-SiH_3$	- 10.5	—	[5]
$^{\cdot}CH_2-CH_2-CHCH_2$	- 10.5	—	[101]
$^{\cdot}CH_2-Si(CH_3)_3$	- 10.9	—	[75]

Table 1 (continued)

Radical	RSE (ROMP2)	RSE (other)	Refs.
$^{\cdot}CH_2-NO_2$	– 11.7	– 11.6 (G3(MP2)-RAD)	[10, 20, 48, 75, 96, 98, 99]
$^{\cdot}CH_2-O-CF_3$	– 11.8	—	[51]
$^{\cdot}CH_2-CH_2-CH_3$	– 11.9	– 13.4 (CBS-4) – 12.5 (G3(MP2)-RAD)	[7, 10, 20, 49, 50, 57, 72, 103]
$^{\cdot}CH_2-F$	– 12.9	– 13.8 (CBS-4) – 12.4 (G3(MP2)-RAD) – 13.0 (W1)	[4, 7, 10, 12, 19, 20, 49, 57, 68–70, 75, 79, 82, 94, 95, 98, 102]
$^{\cdot}CH_2-CH_3$	– 13.8	– 14.1 (G3(MP2)-RAD) – 15.9 (CBS-4)	[4, 5, 7, 10, 18, 20–22, 24, 31, 43, 44, 47–51, 57, 60, 72, 75, 88, 95, 97, 99, 100, 102]
$^{\cdot}CH_2-Br$	– 14.4	—	[7, 20, 94]
$^{\cdot}CH_2-O-CHO$	– 16.4	—	[10]
$^{\cdot}CH_2-O-CO-CH_3$	– 17.7	– 18.5 (G3(MP2)-RAD)	[20]
$^{\cdot}CH_2-CO-O-H$	– 20.0	– 21.2 (G3(MP2)-RAD)	[10, 20, 21, 95, 98]
$^{\cdot}CH_2-CO-NH_2$	– 20.3	—	[10, 23, 60, 75]
$^{\cdot}CH_2-CO-O-CH_3$	– 20.4	– 21.5 (G3(MP2)-RAD)	[20, 75, 102]
$^{\cdot}CH_2-CO-NH-CH_3$	– 20.5	—	[23]
$^{\cdot}CH_2-P(CH_3)_2$	– 20.6	—	
$^{\cdot}CH_2-Cl$	– 20.9	– 21.1 (G3(MP2)-RAD)	[7, 10, 20, 59, 75, 82, 94, 98, 102]
$^{\cdot}CH_2-PH_2$	– 21.5	– 23.3 (G3(MP2)-RAD)	[10, 20, 94]
$^{\cdot}CH_2-CH(CH_2)_2$	– 23.2	—	[72, 101, 103]
$^{\cdot}CH_2-CO-C_6H_5$	– 26.5	—	[74, 75, 96, 102, 104]
$^{\cdot}CH_2-CO-CH_3$	– 28.2	—	[48, 75, 95, 96, 99, 102, 104]

Table 1 (continued)

Radical	RSE (ROMP2)	RSE (other)	Refs.
$^{\cdot}CH_2-CN$	– 31.1	– 31.9 (G3(MP2)-RAD) – 32.7 (W1)	[4, 7, 10, 12, 19–21, 48, 74, 75, 79, 94–96, 98, 99, 102, 105, 106]
$^{\cdot}CH_2-O-CH_3$	– 31.1	– 31.0 (G3(MP2)-RAD)	[7, 10, 20, 51, 58, 75, 94, 99, 102]
$^{\cdot}CH_2-CHO$	– 32.2	– 36.6 (W1) – 34.9 (G3(MP2)-RAD)	[4, 7, 10, 19, 20, 23, 48, 60, 79, 94, 95, 97, 98]
$^{\cdot}CH_2-OH$	– 32.3	– 31.6 (G3(MP2)-RAD) – 33.6 (W1)	[4, 5, 7, 10, 12, 20–22, 31, 48, 58, 75, 79, 94, 95, 97, 98, 102, 106]
$^{\cdot}CH_2-SH$	– 35.2	– 36.1 (G3(MP2)-RAD)	[10, 20, 69, 94, 95, 97, 98]
$^{\cdot}CH_2-S-CH_2-C_6H_5$	– 36.0	– 37.9 (G3(MP2)-RAD)	[13]
$^{\cdot}CH_2-S-CH_3$	– 39.3	– 40.7 (G3(MP2)-RAD)	[14]
$^{\cdot}CH_2-BH_2$	– 40.3	– 40.1 (G3(MP2)-RAD)	[4, 5, 10, 20, 79, 105]
$^{\cdot}CH_2-NH-CHO$	– 41.2	—	[23]
$^{\cdot}CH_2-NH-CO-CH_3$	– 42.0	—	[23]
$^{\cdot}CH_2-NH_2$	– 45.8	– 49.2 (W1) – 44.2 (G3(MP2)-RAD)	[4, 5, 7, 10, 20, 23, 48, 59, 74, 75, 79, 94, 95, 98, 99, 102, 105]
$^{\cdot}CH_2-N(CH_3)_2$	– 45.8	—	[59, 75, 102, 106]
$^{\cdot}CH_2-NH-CH_3$	– 47.1	—	[10, 59, 75, 102]
$^{\cdot}CH_2-C_6H_4-NO_2$	– 49.2	—	[107]
$^{\cdot}CH_2-C_6H_5$	– 50.4	– 58.9 (G3(MP2)-RAD)	[2, 3, 7, 17, 18, 20, 21, 43, 44, 47, 72, 75, 88, 99, 102, 107, 109, 110, 133]
$^{\cdot}CH_2-CCH$	– 50.6	– 52.6 (G3(MP2)-RAD)	[7, 10, 20, 79]
$^{\cdot}CH_2-C_6H_4-CN$	– 50.9	—	[107]
$^{\cdot}CH_2-C_6H_4-OH$	– 52.8	—	[107]

Table 1 (continued)

Radical	RSE (ROMP2)	RSE (other)	Refs.
$^{\cdot}CH_2-C_6H_4-OCH_3$	- 53.2	—	[107]
$^{\cdot}CH_2-CH=CH_2$	- 77.5	- 70.7 (G3(MP2)-RAD)	[3, 4, 7, 10, 19–21, 48, 72, 75, 79, 97, 99, 102]
$^{\cdot}CH_2-CH=CH-CH_3$	- 79.5[a] - 91.3[b]	—	[3, 4, 7]
$^{\cdot}CH_2-CH=C(CH_3)_2$	- 79.0[d] - 91.9[c]	—	[7]

[a] from *trans*-2-butene
[b] from 1-butene
[c] from 3-methyl-1-butene
[d] from 2-methyl-2-butene

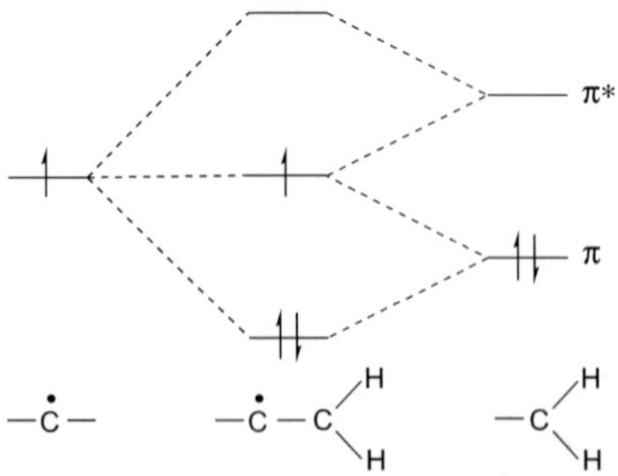

Scheme 2

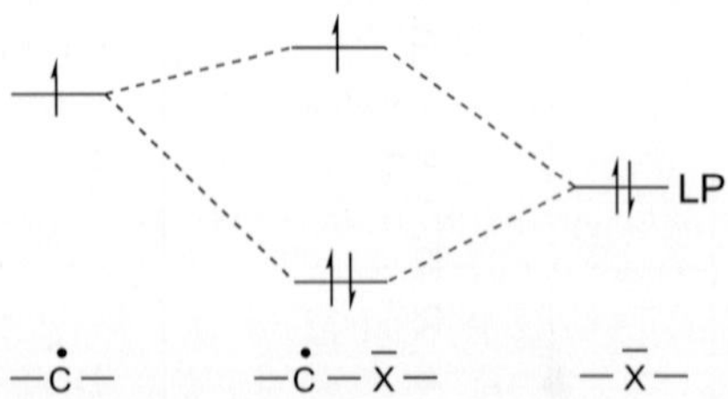

Scheme 3

the C – C double bond. The interaction is, however, less efficient due to the size of this type of π system as well as its energy levels. Moreover, the structure of small alkyl radicals shows that the alignment of the π-type orbital at the radical center with only one neighboring C – H bond appears to be more efficient as compared to the interaction with a π orbital formally constructed from two C – H bonds. The net effect of this type of hyperconjugative interaction is small but stabilizing in any case. The ethyl radical is a typical example for this situation with a stabilization energy of – 13.8 kJ/mol. This stabilizing interaction is reflected in a shortening of the bond connecting the radical center with the alkyl substituent (C – C bond in ethane: 153.1 pm; in ethyl radical: 148.9 pm), but the radical center remains slightly pyramidalized in practically all alkyl radicals. The out-of-plane (oop) bending angle[1] amounts to 9.8° in the ethyl radical and somewhat larger values are found in secondary and tertiary alkyl radicals. Increasing the size of the attached alkyl group does not necessarily lead to more efficient stabilization as can be seen from the values for the 1-propyl radical (– 11.9 kJ/mol) and the *tert*-butylmethyl radical (– 6.5 kJ/mol). However, strained alkyl groups such as cyclopropyl provide π-type orbitals constructed from strained C – C bonds that interact more efficiently with the radical center. This results in a significantly

[1] The out-of-plane (oop) bending angle is in this context defined as the deviation of one of the radical substituents from the plane defined by the radical center and the other two substituents.

larger stabilization energy for the cyclopropylmethyl radical (– 23.2 kJ/mol) as compared to other primary alkyl radicals.

The interaction of radical centers with adjacent lone pair electrons offers a third type of stabilizing interaction. Quite generally this two-center, three-electron interaction is most effective in the presence of high-lying lone pair orbitals. The magnitude of the stabilization energy is largest for amino substituents and smallest for fluorine, indicating a clear dependence on the substituent electronegativity for first row elements. For substituents based on the second row elements P, S, and Cl there is no simple relationship between electronegativity and stabilization energy. A prototype system for this case is the hydroxymethyl radical $HO-CH_2\cdot$ with a stabilization energy of – 32.3 kJ/mol. The C – O bond distances in methanol (141.9 pm) and hydroxymethyl radical (137.0 pm) indicate that the stabilizing interaction also leads to a contraction of the bond between radical center and substituent in this case. The radical center is strongly pyramidalized with an oop-bending angle of 35.6°. It is generally found in these systems that the degree of pyramidalization depends on the number and electronegativity of the substituents. This latter point has been studied repeatedly in fluorinated alkyl radicals, and the series $^{\cdot}CH_2F$, $^{\cdot}CF_2H$, $^{\cdot}CF_3$ with oop-bending angles of 31.4°, 49.5°, and 55.1° (all at the B3LYP/6-31G(d) level) serves as an illustrative example [49, 52–56, 68–70].

The effects of substituents in the β-position to the radical center are mostly inductive in nature. Comparison of the RSE values for the ethyl radical (– 13.8 kJ/mol) with those of the propyl, 2-hydroxyethyl, 2-fluoroethyl, and 2-chloroethyl radicals with RSE values of – 11.9, – 8.5, – 5.5, and – 6.7 kJ/mol also indicates that electronegative substituents in the β-position uniformly destabilize the radical center, the effect being larger for more electronegative substituents. Comparison of the RSE values of the 2-fluoro, 2,2-difluoro, and 2,2,2-trifluoroethyl radicals of – 5.5, + 3.0, and + 8.1 kJ/mol also indicate that these effects can accumulate to yield overall destabilized radicals relative to the methyl radical. Even less favorable RSE values are found for positively charged substituents directly attached to the radical center such as $-NH_3^+$ (+ 18.3 kJ/mol) or $-SH_2^+$ (+ 12.8 kJ/mol) (Table 1).

1.3
Multiply Substituted Alkyl Radicals

The cumulative effects of multiple substituents have been studied at length in search of particularly stable radicals. It is generally found that the repetitive addition of identical substituents leads to a stepwise decrease in RSE values. This is well illustrated by the comparison of the methyl, ethyl, isopropyl, and *tert*-butyl radicals with RSE values of 0.0, – 13.8, – 23.3, and – 28.3 kJ/mol. Thus, while the stability of the alkyl radicals clearly increases with the number of alkyl substituents attached to the radical center, the substituent effects become smaller with increasing stability of the radicals. This is sometimes referred to as a "saturation effect" [71]. This trend can also be observed for resonance-stabilized radicals such as the methyl, benzyl, diphenyl methyl, and triphenyl methyl radicals with RSE values of 0.0, – 50.4, – 78.4, and – 103.4 kJ/mol [47, 72]. Rather extensive work has been published on "captodative" or push/pull-substituted radicals, in which the radical center is connected to one electron-donating and one electron-withdrawing substituent [23, 71, 73–75]. The most prominent systems, in which this effect may be expected, are radicals derived from amino acids and peptides, the glycine-2-yl radical (**9**) being a typical case [23, 60, 63]. The RSE for this radical amounts to – 100.6 kJ/mol (Table 2), which is significantly more than the sum of the RSEs for the aminomethyl radical (– 45.8 kJ/mol) and the carboxymethyl radical (– 20.0 kJ/mol). In a valence bond picture the excess stabilization energy of – 34.8 kJ/mol of the glycine radical can be attributed to zwitterionic Lewis structures such as **9b** and **9c** (Scheme 4), which are present in the combined systems, but absent in the corresponding monosubstituted radicals [23, 73].

One point of debate in defining the magnitude of the captodative effect has been the separation of substituent effects on the radical itself as compared to that on the closed shell reference system. This is, as stated before, a general problem for all definitions of radical stability based on isodesmic reactions such as Eq. 1 [7, 74, 76], but becomes particularly important in multiply substituted cases. This problem can be approached either through estimating the substituent effects for the closed shell parents separately [77, 78], or through the use of isodesmic reactions such as Eq. 5, in which only open shell species are present:

10 **11** **9** **1** (5)

The reaction energy for Eq. 5 amounts to – 54.7 kJ/mol at the ROMP2 level, indicating a substantial captodative stabilization for glycyl radical (**9**) even with this definition.

Radicals derived from hydrofluorocarbons (HFCs) as well as hydrofluoroethers (HFE) are often destabilized with respect to the methyl radical [51, 57, 68, 70, 79–82]. The low stability of these radicals implies that the C – H bonds in the corresponding closed shell parent compounds are comparatively strong and thus rather unreactive towards attack of oxidizing reagents. This latter property is of outstanding importance for the use of these compounds in a variety of technical applications, in which thermally stable, non-oxidizable, non-flammable compounds are needed. However, with respect to the environmental fate of these compounds high C – H bond energies

Table 2 Radical stabilization energies (in kJ/mol) of multiply substituted methyl radicals at 0 K according to Eq. 1

Radical	RSE (ROMP2)	RSE (other)	Refs.
$^{\bullet}CF_3$	+ 10.3	+ 4.2 (CBS-4)	[7, 10, 49, 52–57, 68, 70, 78, 81, 82]
$^{\bullet}CF_2-O-CF_3$	+ 4.7	—	[51]
$^{\bullet}CF_2-CF_3$	- 3.9	- 7.1 (CBS-4)	[43, 49, 51, 57]
$^{\bullet}CClF_2$	- 8.5	—	[82]
$^{\bullet}CHF_2$	- 10.4	- 13.8 (CBS-4)	[7, 10, 49, 57, 68–70, 79, 82, 98]
$^{\bullet}CF_2-CH_3$	- 12.5	- 17.2 (CBS-4)	[85, 93, 94, 134]
$^{\bullet}CHClF$	- 19.8	—	[82]
$^{\bullet}CHF-CH_3$	- 19.9	- 23.4 (CBS-4)	[57]
$^{\bullet}CH(CH_3)_2$	- 23.3	- 27.2 (CBS-4) - 24.0 (G3(MP2)-RAD)	[7, 10, 18, 22, 24, 31, 47, 48, 57, 72, 95, 97, 103]
$^{\bullet}CCl_2F$	- 25.0	—	[82]
$^{\bullet}C(CH_3)_3$	- 28.3	- 34.3 (CBS-4) - 29.7 (G3(MP2)-RAD)	[7, 10, 18, 22, 24, 31, 47, 48, 56, 57, 72, 74, 95, 97]
$^{\bullet}CH(OH)_2$	- 30.4	—	[7, 10]
$^{\bullet}CHCl_2$	- 32.1	—	[7, 10, 82, 98]
$^{\bullet}CCl_2-CH_3$	- 35.6	—	
$CH_3-{}^{\bullet}CH-O-CH_2-CH_3$	- 36.1	—	
$HO-CH_2-{}^{\bullet}CH-OH$	- 36.7	—	
$CH_3-{}^{\bullet}CH-OH$	- 37.5	—	[7, 58, 106]
$(CH_3)_2C^{\bullet}-OH$	- 39.8	—	[7, 106]
$CH_3-{}^{\bullet}CH-CO-OCH_3$	- 40.3	—	

Table 2 (continued)

Radical	RSE (ROMP2)	RSE (other)	Refs.
$^{\cdot}CCl_3$	– 42.4	—	[7, 10, 54–56, 82, 95]
$NH_2-{}^{\cdot}CH-CH_3$	– 49.8	—	[59]
$NH_2-{}^{\cdot}C(CH_3)_2$	– 51.8	—	[59]
$^{\cdot}C(CH_3)(SCH_3)_2$	– 52.2	– 59.9 (G3(MP2)-RAD)	[14]
$(CH_3)_2C^{\cdot}-CO-OCH_3$	– 53.9	—	
$CH_3-{}^{\cdot}CH-CHO$	– 54.1	—	[48]
$^{\cdot}C(CH_3)_2CN$	– 59.4	– 59.0 (G3(MP2)-RAD)	[7, 13]
$^{\cdot}CH(CH_3)Ph$	– 59.7	—	[7, 18, 47, 72]
$^{\cdot}C(CH_3)_2Ph$	– 62.1	—	[7, 18, 47, 72]
$^{\cdot}CH(CN)_2$	– 66.0	—	[7, 71, 98]
$NC-{}^{\cdot}CH-OH$	– 73.0	—	[7]
$C_6H_5-{}^{\cdot}CH(CO-C_6H_5)$	– 73.6	—	[104]
$HCO-NH-{}^{\cdot}CH-CO-NH_2$	– 76.0	—	[23, 60, 61]
$CH_3CO-NH-{}^{\cdot}CH-CO-NH-CH_3$	– 77.7	—	[23, 60]
$^{\cdot}CHPh_2$	– 78.4	—	[18, 47, 72, 75, 112, 113]
$NH_2-{}^{\cdot}CH-CN$	– 89.1	—	[7, 71, 74, 79, 98, 105]
$NH_2-{}^{\cdot}CH-CO-NH_2$	– 90.3	—	[23, 60]
$HO-{}^{\cdot}CPh_2$	– 99.2	—	[7]
$NH_2-{}^{\cdot}CH-COOH$	– 100.6	—	[60–66]
$HCO-NH-{}^{\cdot}CH-CHO$	– 102.5	—	[23, 60]
$^{\cdot}C(CN)_3$	– 103.0	—	[7]

Table 2 (continued)

Radical	RSE (ROMP2)	RSE (other)	Refs.
$^{\cdot}CPh_3$	– 103.4	—	[18, 72, 114]
$CH_2CH-{}^{\cdot}CH-OH$	– 104.5 [d]	—	[4, 7, 79]
$NH_2-{}^{\cdot}C(CH_3)-COOH$	– 107.2	—	[63, 67]
$(CH_3)_2N-{}^{\cdot}CH-CO-C_6H_5$	– 117.2	—	[77]
$^{\cdot}CH(CHCH_2)_2$	– 119.9 [a] – 91.0 [b]	—	[72, 79, 106]
$NH_2-{}^{\cdot}CH-CHO$	– 126.2	– 123.0 (W1)	[23, 48, 60, 79, 98]
$H_2N-{}^{\cdot}C(CN)_2$	– 127.1	—	[73]
$NH_2-{}^{\cdot}C(CH_3)-CHO$	– 135.8	—	[48]
$(CH_3)_2N-{}^{\cdot}C(CN)_2$	– 135.8	—	[73]
$NH_2-{}^{\cdot}CH-BH_2$	– 150.7 [c]	—	[79, 105]

[a] from 1,4-pentadiene
[b] from *trans*-1,3-pentadiene
[c] acyclic isomer
[d] from allyl alcohol

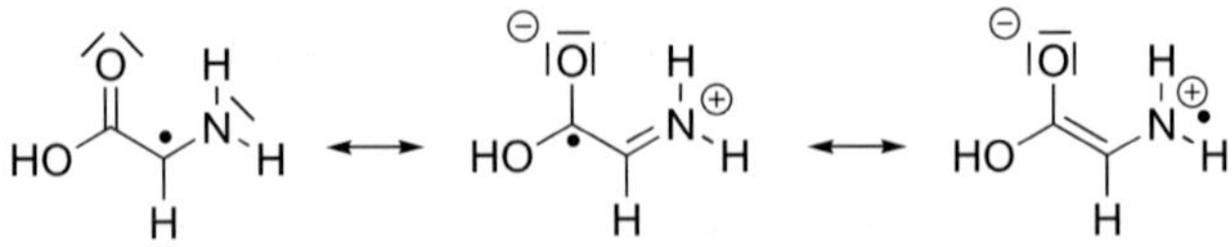

9a 9b 9c

Scheme 4

translate into long atmospheric lifetimes, a property which is not necessarily desirable in replacements of chlorofluorocarbons (CFCs). As is readily seen for the stability values for the methyl, fluoromethyl, difluoromethyl and trifluoromethyl radicals (0.0, – 12.9, – 10.4, and + 10.3 kJ/mol), the substituent effects of fluorine appear to depend on the number and character of other α-substituents present in the radical. Together with the remarkable effect of α-fluorine substituents on the pyramidalization of the radical center this has been taken to reflect the dual nature of fluorine as a σ-acceptor and π-donor substituent [57, 68, 70]. However, as already stated above in the context of the captodative effect, the definition of radical stability according to Eq. 1 also includes substituent effects on the closed shell parents. For multiply fluorinated closed shell compounds one must expect a significant anomeric effect as defined, for example, through the isodesmic reaction in Eq. 6:

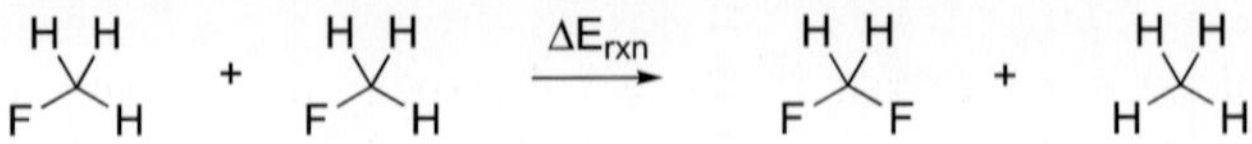

12 12 13 2 (6)

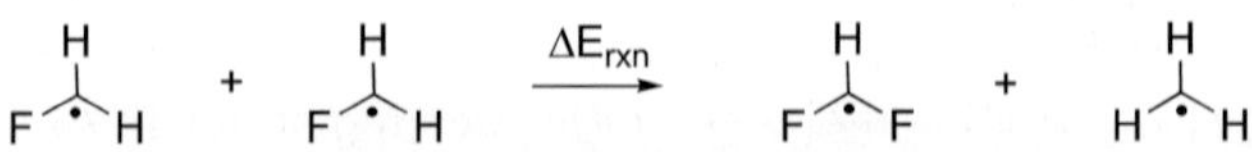

14 14 15 1 (7)

The formal reaction of two molecules of fluoromethane (**12**) to difluoromethane (**13**) and methane (**2**) is strongly exothermic with $\Delta E_{\text{rxn}} =$ – 56.3 kJ/mol (ROMP2 value). This reaction energy is practically identical to that obtained from the corresponding standard heats of formation for these three species [83] (ΔH_{rxn} = – 56.9 kJ/mol) and implies a strongly stabilizing anomeric effect between the two C – F bonds in **13**. Taking the same approach to estimating the anomeric effect in the corresponding radicals with isodesmic reaction Eq. 7 we obtain a value of – 40.8 kJ/mol (ROMP2 value). This indicates that the simultaneous presence of two α-fluorine substituents in radical **15** is indeed stabilizing with respect to the singly fluorinated radicals **14**, but that the degree of stabilization is smaller as compared to the corresponding closed shell compounds. As long as the latter are used as reference systems (as is done in Eq. 1), introduction of a second fluorine substituent at the radical center will indeed be destabilizing.

Table 3 Radical stabilization energies (in kJ/mol) of cyclic alkyl radicals at 0 K according to Eq. 1

Radical	RSE (ROMP2)	RSE (other)	Refs.
$^{\cdot}CH(CH_2)_2$	+ 21.9	—	[7, 9, 72, 103]
$^{\cdot}CH(CH_2)_3$	- 15.6	—	[72, 103, 134]
$^{\cdot}CH(CH_2)_5$	- 19.4	—	[7, 103, 134]
$^{\cdot}CH(CH_2)_4$	- 31.4	—	[72, 103]
1,3-dioxolan-2-yl	- 41.1	—	[58]
tetrahydrofuran-2-yl	- 41.7	—	[58, 106, 112]
pyrrolidine-2-yl	- 55.2	—	[59]
9-fluorenyl	- 90.7	—	[71, 113]
cyclopentene-3-yl	- 93.4	- 95.0	[112, 118]
cyclopentadienyl	- 98.3	—	[7, 112]
cyclohexa-2,4-dien-1-yl	- 124.9 [a] - 120.9 [b]	—	[7, 106, 112, 115–117]
tropylium	- 134.1	—	[133]

[a] from 1,4-cyclohexadiene
[b] from 1,3-cyclohexadiene

In the cyclic radicals summarized in Table 3 the substituent effects are modified through the more or less strained ring systems. Ring strain appears to be quite significant in determining the stability of the cyclopropyl radical (**19**) (RSE = + 21.9 kJ/mol) and cyclobutyl radical (RSE = – 15.6 kJ/mol), both of which are less stable than comparable acyclic alkyl radicals such as the isopropyl radical (**20**) (RSE = – 23.3 kJ/mol). The ring strain in cyclopropane (**16**) has been assessed with the homodesmotic reaction described in Eq. 8 [84, 85]. Homodesmotic reactions represent a subgroup of isodesmic reactions, in which the number of bonds between centers of identical character are conserved together with their bonding environment. Using energies calculated at the ROMP2 level a strain energy of + 111.9 kJ/mol is obtained for **16**.

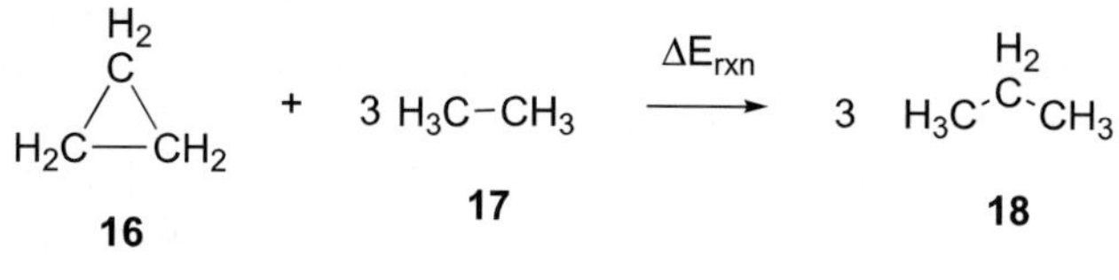

(8)

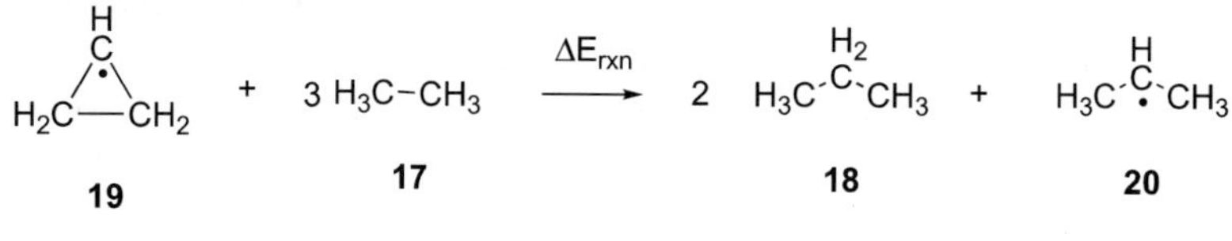

(9)

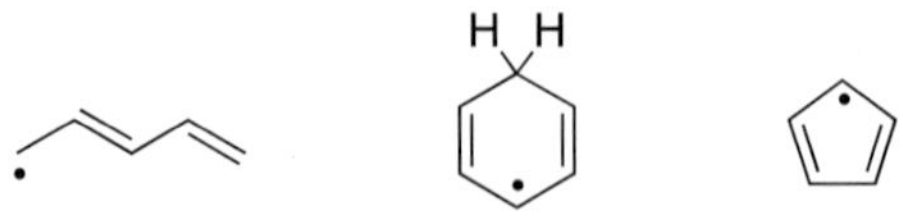

21 22 23

Scheme 5

Taking an analogous approach for the cyclopropyl radical (**19**) as expressed in Eq. 9, a strain energy value of + 157.1 kJ/mol is obtained. The difference of 45.2 kJ/mol indicates that formation of a radical center in the cyclopropane ring increases the ring strain quite significantly. This is most easily explained as the consequence of significantly different C – C – C bond angles in propane (**18**) with 112.9° and isopropyl radical (**20**) with 121.5°. On formal cyclization to cyclopropane (**16**) and the cyclopropyl radical (**19**) these angles have to be compressed to 60.0° and 63.1°, respectively. For larger cycloalkyl radicals the effects of ring strain become less evident. For heterocyclic radicals it appears that stabilization of the radical center through adjacent lone pairs (e.g. in the tetrahydrofuran-2-yl radical) is as efficient as in the closest acyclic analogs. One interesting ring size effect is visible in the pentadienyl radicals, in which the acyclic 2,4-pentadien-1-yl radical (**21**) and the cyclohexa-2,4-dien-1-yl radical (**22**) have very similar stabilities, while the corresponding five-membered ring form [cyclopentadienyl radical (**23**)] is significantly less stable (Table 3).

1.4 The Stability of Delocalized Radicals

When considering the stability of spin-delocalized radicals the use of isodesmic reaction Eq. 1 presents one further problem, which can be illustrated using the 1-methyl allyl radical **24**. The description of this radical through resonance structures **24a** and **24b** indicates that **24** may formally be considered to either be a methyl-substituted allyl radical or a methylvinyl-substituted methyl radical. While this discussion is rather pointless for a delocalized, resonance-stabilized radical such as **24**, there are indeed two options for the localized closed shell reference compound. When selecting 1-butene (**25**) as the closed shell parent, C – H abstraction at the C3 position leads to **24** with a radical stabilization energy of – 91.3 kJ/mol, while C – H abstraction from the C1 position of *trans*-2-butene (**26**) generates the same radical with a RSE value of – 79.5 kJ/mol (Scheme 6). The difference between these two values (12 kJ/mol) reflects nothing else but the stability difference of the two parents **25** and **26**.

Similarly, the resonance stabilization energy of radical **21** may either be defined relative to 1,4-pentadiene (**27**) (RSE = – 119.9 kJ/mol) or to *trans*-1,3-pentadiene (**28**) (RSE = – 91.0 kJ/mol). These examples indicate that the RSE

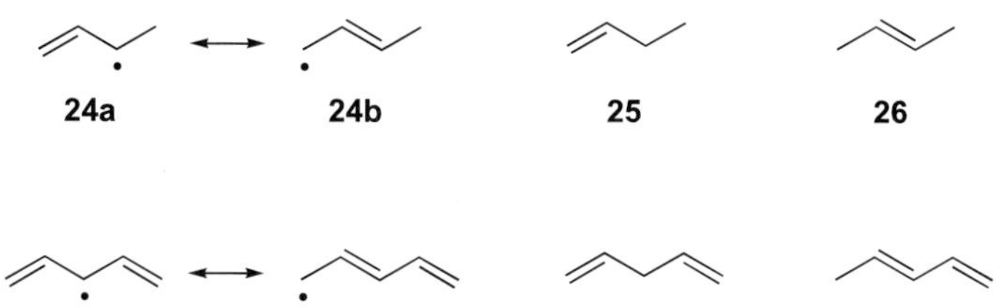

24a 24b 25 26

21a 21b 27 28

Scheme 6

values collected in Tables 1–6 are only meaningful with respect to one clearly defined closed shell reference compound. Where this is not given explicitly in the literature, it is generally assumed that the RSE value refers to the closed shell compound obtained from the radical "as drawn" through addition of a hydrogen atom.

1.5 The Stability of σ-Alkyl Radicals

The substituent effects predicted for vinyl radicals are rather similar to those already observed for alkyl radicals (Table 4). Attachment of alkyl groups or π systems to the radical center stabilize the radical while the introduction of σ-acceptors in the α- or β-position are destabilizing. The nature of the

Table 4 Radical stabilization energies (in kJ/mol) of σ-radicals at 0 K according to Eq. 2

Radical	RSE (RMP2)	RSE (other)	Refs.
$F-{}^{\bullet}CCF_2$	+32.2	+30.1 (CBS-4)	[57]
$H-{}^{\bullet}CCF_2$	+31.1	+31.8 (CBS-4)	[57]
$CF_3-{}^{\bullet}CCH_2$	+13.1	+13.0 (CBS-4)	[57]
${}^{\bullet}C_6H_5$	+10.3	—	[7, 17, 97]
$F-{}^{\bullet}CCH_2$	+7.2	+6.3 (CBS-4) +8.2 (Martin-3)	[19, 57]
$H-{}^{\bullet}CCH_2$	0.0	0.0	[7, 21, 22, 31, 57, 97, 103, 111, 118]
$CH_3-{}^{\bullet}CCH_2$	- 11.9	- 13.4 (CBS-4)	[118]
$NC-{}^{\bullet}CCH_2$	- 19.5	- 17.1 (Martin-3)	[19, 119]
$C_6H_5-{}^{\bullet}CCH_2$	- 28.3	—	
${}^{\bullet}C(O)OH$	- 57.6	—	[97]
${}^{\bullet}CO(CH_3)$	- 98.1	- 90.4 (CBS-4) - 90.8 (G2)	[21, 22, 31, 35, 97, 99]
${}^{\bullet}HCO$	- 104.5	- 90.8 (CBS-4) - 94.6 (G2)	[21, 22, 31, 35, 97]

α-substituent also has a large influence on the structure of σ-radicals: while the vinyl radical features a H – C – C bond angle of 137.5°, the 1-phenylvinyl radical is perfectly linear at the radical center. Particularly stable σ-radicals are obtained from carbonyl compounds, in which the carbonyl lone pair electrons are effectively stabilizing neighbors of the radical center.

2 Heteroatom-Based Radicals

2.1 Nitrogen-Centered Radicals

Nitrogen-centered radicals have been studied thoroughly in the context of radicals derived from amino acids and peptides [23, 40]. The substituent effects predicted for amino radicals through Eq. 3 clearly reflect that nitrogen-centered radicals are more electron-deficient species as compared to the corresponding carbon analogs (Table 5). The electron donating ability of simple alkyl groups is stabilizing in both cases but the effect is much larger in the methylamino radical (RSE = – 29.4 kJ/mol) as compared to the ethyl radical (RSE = – 13.8 kJ/mol). Addition of a second alkyl group as in the *N,N*-dimethylamino radical (RSE = – 52.1 kJ/mol) is also much more significant as in the isopropyl radical (RSE = – 23.3 kJ/mol). The attachment of larger π systems as in the phenylamino and the diphenylamino radicals (RSE = – 49.2 and – 82.2 kJ/mol) leads to stabilization of the radical center in much the same way as already discussed for carbon-centered radicals, the stabilization energies also being of comparable magnitude. However, in clear contrast to alkyl radicals, the attachment of carbonyl groups to the amino radical center is destabilizing in all cases. This is a consequence of resonant delocalization of the unpaired spin into the π system of the carbonyl group, which is accompanied by the loss of resonant interaction between the carbonyl group and the nitrogen lone pair [23, 40, 41]. Comparison of the RSE values calculated at different levels of theory indicates that this demanding situation is not very-well described with economical methods such as ROMP2 and that a reliable prediction can in this case only be expected from compound methods such as CBS-QB3 or W1 [23, 40].

2.2 Oxygen-Centered Radicals

The stability of oxygen-centered radicals has been studied repeatedly in recent years in search of quantitative descriptors of antioxidant activity. The antioxidant activity of phenols is indeed so well correlated with O – H BDEs and the ionization potential that these two energies can be used as the guiding

Table 5 Radical stabilization energies (in kJ/mol) of nitrogen-centered radicals at 0 K according to Eq. 3

Radical	RSE (ROMP2)	RSE (other)	Refs.
$^{\cdot}NH-CHO$	+43.5	+30.1 (W1) +26.0 (CBS-QB3)	[23, 40, 41]
$^{\cdot}NH-CO-CH_3$	+35.4	+19.0 (CBS-QB3)	[23, 40, 41]
$CHO-CH_2-{}^{\cdot}N-CHO$	+28.3	+16.9 (CBS-QB3)	[23]
$^{\cdot}NH-CO-CH_2-NH_2$	+28.1	+20.9 (CBS-QB3)	[23]
$NH_2-CO-CH_2-{}^{\cdot}N-CO-CH_3$	+19.0	+11.5 (CBS-QB3)	[23]
$CH_3-{}^{\cdot}N-CHO$	+18.4	+3.5 (CBS-QB3)	[23]
$CHO-CH_2-{}^{\cdot}N-CO-CH_3$	+18.0	+11.2 (CBS-QB3)	[23]
$^{\cdot}NH-CO-NH_2$	+14.1	+7.4 (W1) +5.4 (CBS-QB3)	[40]
$^{\cdot}NH-CF_3$	+12.5	+11.1 (W1) +11.6 (CBS-QB3)	[40, 41]
$CH_3-{}^{\cdot}N-CO-CH_3$	+10.6	– 1.8 (CBS-QB3)	[23]
$^{\cdot}NH_2$	0.0	0.0	[7, 23, 40, 41, 44, 97, 102]
$^{\cdot}NH-CH_2-CHO$	– 23.4	– 26.0 (CBS-QB3)	[23]
$^{\cdot}NH-CH_2-CO-NH_2$	– 26.6	– 28.7 (CBS-QB3)	[23]
$^{\cdot}NH-CH_3$	– 29.4	– 32.0 (W1) – 31.8 (CBS-QB3)	[7, 23, 41, 72, 100]
$^{\cdot}NH-C_6H_5$	– 49.2	—	[7, 102, 106, 108, 109, 120–123]
$^{\cdot}N(CH_3)_2$	– 52.1	—	[7]
$^{\cdot}NPh_2$	– 82.2	—	[7, 102, 106, 120, 123, 124]

Table 6 Radical stabilization energies (in kJ/mol) of oxygen-centered radicals at 0 K according to Eq. 4

Radical	RSE (ROMP2)	RSE (other)	Refs.
$^{\bullet}O-CF_3$	+6.6	—	[42, 125]
$^{\bullet}O-H$	0.0	—	[7, 21, 22, 44]
$^{\bullet}O-C(CH_3)_3$	- 42.5	—	[7, 17, 100, 125]
$^{\bullet}O-CH_2-CH_3$	- 50.9	—	[22, 125, 134, 135]
$^{\bullet}O-CH_3$	- 52.3	- 48.1 (CBS-4) - 54.0 (G2)	[7, 21, 22, 35, 44, 100, 125, 135]
$^{\bullet}O-OH$	- 125.0	—	[21, 42, 97, 135]
$^{\bullet}O-C_6H_4-NO_2$ (para)	- 140.6	—	[42, 123, 126–129]
$^{\bullet}O-C_6H_5$	- 149.8	—	[7, 42, 88, 106, 108, 109, 121] [123, 126–132]
$^{\bullet}O-NH_2$	- 177.1	—	[42]
$^{\bullet}O-C_6H_4-NH_2$ (para)	- 197.3	—	[42, 123, 126–130]

principles in developing improved antioxidants [45, 46, 86–91]. For smaller alkoxy radicals there is, unfortunately, much less quantitative data available. For a radical centered at the electronegative oxygen atom one must expect substituent effects similar to those for nitrogen-centered radicals. Comparison of the RSE values for the ethyl, aminomethyl and methoxy radicals of - 13.8, - 29.4 and - 52.3 kJ/mol clearly illustrates that electron donating substituents are highly relevant for the stability of oxygen-centered radicals. This is also underlined by the RSE values for *para*-substituted phenoxy radicals, in which the *para*-aminophenoxy radical is significantly more stable than the *para*-nitrophenoxy radical. Efficient electron donation is also possible through lone pairs located in direct neighborhood to the radical center. This effect makes peroxy radicals much more stable than the analogous alkoxy radicals and is the basis for the remarkable stability of nitroxy radicals.

3 Connecting the Scales

While Eqs. 1–4 may represent the most consistent approach for the definition of radical stability for a selected subclass of radicals, there may still be the need (or desire) to compare the stabilities of radicals characterized through different reference systems. The following three hydrogen transfer reactions can be used to connect the four scales used in this overview:

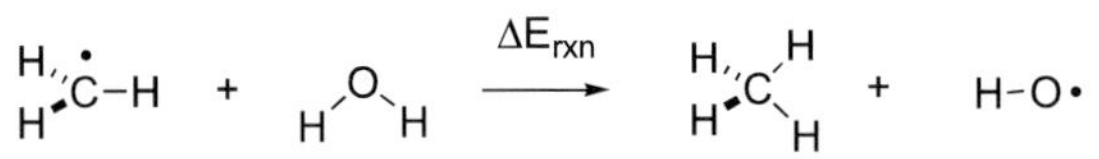

1 8 2 7 (10)

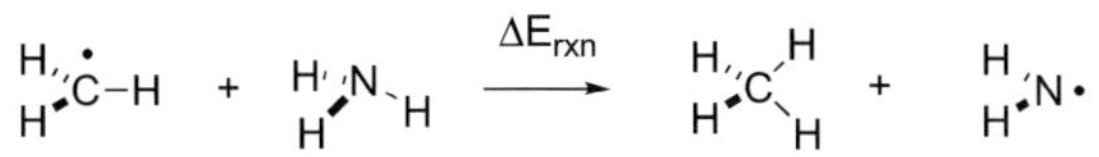

1 6 2 5 (11)

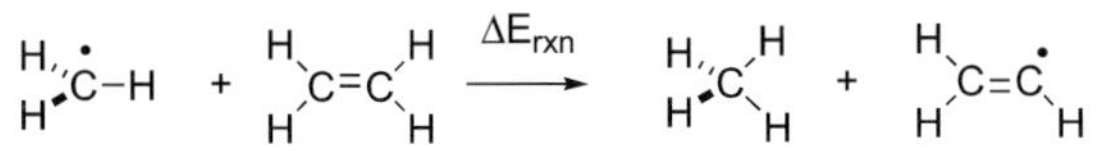

1 4 2 3 (12)

The reaction energies at either 0 or 298.15 K for these reactions as compiled in Table 7 have been assembled from theoretically calculated or experimentally measured heats of formation[2]. Alternatively, the reaction energies can also be calculated from experimentally measured X – H bond dissociation energies for **2**, **4**, **6**, and **8** at either 0 or 298 K [92]. From the results for Eq. 10 we can see that oxygen-centered radicals are less stable than alkyl radicals by close to 60 kJ/mol at 298.15 K. This value is closely reproduced by W1 and W2 theory. In combination with the G3 temperature correction of + 2.9 kJ/mol this indicates that the stabilities of alkyl and alkoxy radicals differ systematically by around 63 kJ/mol at 0 K. On the basis of experimentally measured bond dissociation energies a value of +60.8 is predicted at 0 K. This latter value indicates that the G3 temperature correction may be somewhat too large. The stability differences between the methyl, amino, and vinyl radicals are, in comparison, somewhat smaller. Unfortunately, the rather large uncertainty of the experimentally measured heats of formation for the amino radical (**5**) and the vinyl radical (**3**) makes it difficult to provide a precise value for the stability differences in these two cases. The bond dissociation energy data at 0 K for ammonia and ethylene are, however, known somewhat more accurately and we can thus predict reaction energies of + 14.2 kJ/mol for Eq. 11 and of + 26.8 kJ/mol for Eq. 12 at 0 K. These results are closely matched by combinations of W1 reaction energies at 298 K and the G3 temperature corrections. With these three energy values in hand one can predict how, for example, the stability of the *tert*-butyl radical relates to that of the phenylamino radical. The RSE value for the former amounts to – 28.3 kJ/mol (Table 2) while the RSE value for the latter is – 49.2 kJ/mol (Table 5). In combination with the offset of + 14.2 kJ/mol as the systematic stability difference of alkyl and amino radicals this yields a final stability difference of 6.7 kJ/mol. That is, the *tert*-butyl radical is *less* stable then the phenylamino radical by

[2] All G3 energies have been taken from [15] and the corresponding internet site by Larry A. Curtiss at http://chemistry.anl.gov/compmat/g3theory.htm; all other data have been taken from [39].

Table 7 Reaction enthalpies (in kJ/mol) for hydrogen-transfer reactions (Eqs. 10–12)

Energies	Reaction enthalpy (kJ/mol)
10	
ΔH_f(298),exp.[a]	+ 59.8 ± 1.0
ΔH_f(298),W2[b]	+ 58.6
ΔH_f(298),W1[b]	+ 59.3
ΔH_f(298),G3[a]	+ 57.3
ΔH_f(0),G3[a]	+ 60.2
BDE(0),exp.[c]	+ 61.8 ± 0.6
11	
ΔH_f(298),exp.[a]	+ 13.4 ± 7.4
ΔH_f(298),W2[b]	+ 10.8
ΔH_f(298),W1[b]	+ 10.7
ΔH_f(298),G3[a]	+ 10.5
ΔH_f(0),G3[a]	+ 12.5
BDE(0),exp.[c]	+ 14.2 ± 1.7
12	
ΔH_f(298),exp.[a]	+ 26.0 ± 6.3
ΔH_f(298),W1[b]	+ 24.1
ΔH_f(298),G3[a]	+ 25.1
ΔH_f(0),G3[a]	+ 26.4
BDE(0),exp.[c]	+ 26.8 ± 3.8

[a] All G3 energies have been taken from [15] and the corresponding internet site by Larry A. Curtiss at http://chemistry.anl.gov/compmat/g3theory.htm; all other data have been taken from [39];
[b] [39];
[c] [92]

6.7 kJ/mol. This also implies that hydrogen atom transfer from aniline to the *tert*-butyl radical is exothermic by the same amount of energy.

References

1. Giese B (1989) In: Regitz M, Giese B (eds) C-Radikale, 4th Ed, vol E19a. Houben-Weyl, Thieme, Stuttgart
2. Szwarc M (1948) J Chem Phys 16:128
3. Rodgers AS, Wu MCR, Kuitu C (1972) J Phys Chem 76:918
4. Lehd M, Jensen F (1991) J Org Chem 56:884
5. Coolidge MB, Borden WT (1988) J Am Chem Soc 110:2298
6. Peeters D, Leroy G (1995) Chem Phys Lett 246:481
7. Leroy G (1985) Adv Quant Chem 17:1
8. Pratt DA, DiLabio GA, Mulder P, Ingold KU (2004) Acc Chem Res 37:334

9. Lien MH, Hopkinson AC (1985) J Comp Chem 6:274
10. Pasto DJ, Krasnansky R, Zercher C (1987) J Org Chem 52:3062
11. Bauschlicher CW, Partridge H (1995) J Chem Phys 103:1788
12. Coote ML (2004) J Phys Chem A 108:3865
13. Coote ML (2004) Macromolecules 37:5023
14. Coote ML, Henry DJ (2005) Macromolecules 38:5774
15. Mebel AM, Morokuma K, Lin MC (1995) J Chem Phys 103:7414
16. Froese RDJ, Humbel S, Svensson M, Morokuma K (1997) J Phys Chem A 101:227
17. Froese RDJ, Morokuma K (1999) J Phys Chem A 103:4580
18. Vreven T, Morokuma K (1999) J Chem Phys 111:8799
19. Parkinson CJ, Mayer PM, Radom L (1999) J Chem Soc Perkin Trans 2, 2305
20. Henry DJ, Parkinson CJ, Mayer PM, Radom L (2001) J Phys Chem A 105:6750
21. Henry DJ, Parkinson CJ, Radom L (2002) J Phys Chem A 106:7927
22. Henry DJ, Sullivan MB, Radom L (2002) J Chem Phys 118:4849
23. Wood GPF, Moran D, Jacob R, Radom L (2005) J Phys Chem A 109:6318
24. Izgorodina EI, Coote ML, Radom L (2005) J Phys Chem A 109:7558
25. Scott AP, Radom L (1996) J Phys Chem 100:16502
26. Curtiss LA, Raghavachari K, Trucks GW, Pople JA (1991) J Chem Phys 94:7221
27. Curtiss LA, Carpenter JE, Raghavachari K, Pople JA (1992) J Chem Phys 96:9030
28. Curtiss LA, Raghavachari K, Pople JA (1993) J Chem Phys 98:1293
29. Glukhovtsev M, Pross A, McGrath M, Radom L (1995) J Chem Phys 103:1878
30. Curtiss LA, Redfern PC, Smith BJ, Radom L (1996) J Chem Phys 104:5148
31. Curtiss LA, Raghavachari K, Redfern PC, Rassolov V, Pople JA (1998) J Chem Phys 109:7764
32. Baboul AG, Curtiss LA, Redfern PC, Raghavachari K (1999) J Chem Phys 110:7650
33. Curtiss LA, Redfern PC, Raghavachari K, Rassolov V, Pople JA (1999) J Chem Phys 110:4703
34. Curtiss LA, Redfern PC, Raghavachari K, Pople JA (2001) J Chem Phys 114:108
35. Ochterski JW, Petersson GA, Wiberg KB (1995) J Am Chem Soc 117:11299
36. Ochterski JW, Petersson GA, Montgomery JA Jr (1996) J Chem Phys 104:2598
37. Martin JML, de Oliveira F (1999) J Chem Phys 111:1843
38. Martin JML (1999) Chem Phys Lett 310:271
39. Parthiban S, Martin JML (2001) J Chem Phys 114:6014
40. Wood GP, Henry DJ, Radom L (2003) J Phys Chem A 107:7985
41. Song KS, Cheng YH, Fu Y, Liu L, Li XS, Guo QX (2002) J Phys Chem A 106:6651
42. Fu Y, Liu L, Yi M, Lin BL, Guo QX (2004) THEOCHEM 674:241
43. DiLabio GA, Pratt DA (2000) J Phys Chem A 104:1938
44. DiLabio GA, Pratt DA, LoFaro AD, Wright JS (1999) J Phys Chem A 103:1653
45. Wright JS, Carpenter DJ, McKay DJ, Ingold KU (1997) J Am Chem Soc 119:4245
46. Wright JS, Johnson ER, DiLabio GA (2001) J Am Chem Soc 123:1173
47. Vreven T, Morokuma K (1999) J Chem Phys 111:8799
48. Brinck T, Lee HN, Jonsson M (1999) J Phys Chem A 103:7094
49. Bartberger MD, Dolbier WR Jr, Lusztyk J, Ingold KU (1997) Tetrahedron 53:9857
50. White JC, Cave RJ, Davidson ER (1088) J Am Chem Soc 110:6308
51. Lazarou YG, Papagiannakopoulos P (1999) Chem Phys Lett 301:19
52. Fukuaya H, Ono T, Abe T (1998) J Comp Chem 19:277
53. Guerra M (1995) J Phys Chem 99:81
54. Pakiari AH, Nazari F (2005) THEOCHEM 717:189
55. Cheong BS, Cho HG (1997) J Phys Chem A 101:7901
56. Guerra M (1995) Pure Appl Chem 67:797

57. Zhang XM (1998) J Org Chem 63:3590
58. Shtarev AB, Tian F, Dolbier WR Jr, Smart BE (1999) J Am Chem Soc 121:7335
59. Wayner DDM, Clark KB, Rauk A, Armstrong DA (1997) J Am Chem Soc 119:8925
60. Himo F (2001) Chem Phys Lett 328:270
61. Rauk A, Yu D, Taylor J, Shustov GV, Block DA, Armstrong DA (1999) Biochemistry 38:9089
62. Armstrong DA, Yu D, Rauk A (1996) Can J Chem 74:1192
63. Jonsson M, Wayner DDM, Armstrong DA, Yu D, Rauk A (1998) J Chem Soc Perkin Trans 2 1967
64. Yu D, Rauk A, Armstrong DA (1995) J Am Chem Soc 117:1789
65. Leroy G, Sana M, Wilante C (1991) J Mol Struct 228:37
66. Block DA, Yu D, Armstrong DA, Rauk A (1998) Can J Chem 76:1042
67. Rauk A, Yu D, Armstrong DA (1997) J Am Chem Soc 119:208
68. Bernardi F, Cherry W, Shaik S, Epiotis ND (1978) J Am Chem Soc 100:1352
69. Cramer CJ (1991) J Org Chem 56:5229
70. Bernardi F, Epiotis ND, Cherry W, Schlegel HB, Whangbo MH, Wolfe S (1976) J Am Chem Soc 198:469
71. Bordwell FG, Lynch TY (1989) J Am Chem Soc 111:7558
72. Fattahi A, Kass SR (2004) J Org Chem 69:9176
73. Viehe HG, Janousek Z, Mereny R, Stella L (1985) Acc Chem Res 18:148
74. Sustmann R, Korth HG (1990) Adv Phys Org Chem 26:131
75. Bordwell FG, Zhang XM, Alnajjar MS (1992) J Am Chem Soc 114:7623
76. Coote ML, Pross Am Radom L (2003) Org Lett 5:4689
77. Welle FM, Beckhaus HD, Rüchardt C (1997) J Org Chem 62:552
78. Rüchardt C, Beckhaus HD (1985) Top Curr Chem 130:1
79. Pasto DJ (1988) J Am Chem Soc 110:8164
80. Dolbier WR Jr (1996) Chem Rev 96:1557
81. Schneider WF, Wallington TJ (1993) J Phys Chem 97:12783
82. Luke BT, Lowe GH, McLean AD (1987) J Am Chem Soc 109:1307
83. NIST Standard Reference Database No. 69, 6/2005 release
84. Alcami M, Mo O, Yanez M (1998) J Comp Chem 19:1072
85. Bach RD, Dmitrenko O (2002) J Org Chem 67:2588
86. Wijtmans M, Pratt DA, Valgimigli L, DiLabio GA, Pedulli GF, Porter NA (2003) Angew Chem 115:4506
87. Wijtmans M, Pratt DA, Valgimigli L, DiLabio GA, Pedulli GF, Porter NA (2003) Angew Chem Int Ed 42:4370
88. Mulder P, Korth HG, Pratt DA, DiLabio G, Valgimigli L, Pedulli GF, Ingold KU (2005) J Phys Chem A 109:2647
89. Hussain HH, Babic G, Durst T, Wright JS, Flueraru M, Chichirau A, Chepelev LL (2003) J Org Chem 68:7023
90. Valgimigli L, Brigati G, Pedulli GF, DiLabio GA, Mastragostino M, Arbizzani C, Pratt DA (2003) Chem Eur J 9:4997
91. Wijtmans M, Pratt DA, Valgimigli L, Serwa R, Brinkhorst J, Pedulli GF, Porter NA (2004) J Org Chem 69:9215
92. Berkowitz J, Ellison GB, Gutman D (1994) J Phys Chem 98:2744
93. Song KS, Liu L, Guo QX (2003) J Org Chem 68:4604
94. Mayer PM, Glukhovtsev MN, Gauld JM, Radom L (1997) J Am Chem Soc 119:12889
95. Korth HG, Sicking W (1997) J Chem Soc Perkin Trans 2:715
96. Bordwell FG, Harrelson JA, Zhang XM (1991) J Org Chem 56:4448
97. Blanksby SJ, Ellison GB (2003) Acc Chem Res 36:255

98. Leroy G, Dewispelaere JP, Benkadour H, Temsamani DR, Wilante C (1994) Bull Soc Chim Belg 103:367
99. Jursic BS, Timberlake JW, Engel PS (1996) Tetrahedron Lett 37:6473
100. Fu Y, Liu L, Liu R, Guo QX (2005) J Phys Org Chem 18:529
101. Martinez FN, Schlegel HB, Newcomb M (1998) J Org Chem 63:3618
102. Bordwell FG, Zhang XM (1993) Acc Chem Res 26:510
103. Bach RD, Dmitreno O (2004) J Am Chem Soc 124:4444
104. Alnajjar MS, Zhang XM, Gleicher GJ, Truksa SV, Franz JA (1992) J Org Chem 67:9016
105. Crans D, Clark T, Schleyer PVR (1980) Tetrahedron Lett 21:3681
106. Laarhoven LJJ, Mulder P, Wayner DDM (1999) Acc Chem Res 32:342
107. Pratt DA, Wright JS, Ingold KU (1999) J Am Chem Soc 121:4877
108. Bean GP (2002) Tetrahedron 58:9941
109. Nau WM (1997) J Phys Org Chem 10:445
110. Wu YD, Wong CL, Chan KWK, Jiang XK (1996) J Org Chem 61:746
111. Tian F, Bartberger MD, Dolbier WR Jr (1999) J Org Chem 64:540
112. Laarhoven LJJ, Mulder P (1997) J Phys Chem B 101:73
113. Bordwell FG, Zhang XM, Cheng JP (1991) J Org Chem 56:3216
114. Vreven T, Morokuma K (2002) J Phys Chem A 106:6167
115. Berho F, Rayez MT, Lesclaux R (1999) J Phys Chem A 103:5501
116. Birch AJ, Hinde AL, Radom L (1980) J Am Chem Soc 102:4074
117. Morokuma K, Ohnishi S, Masuda T, Fukui K (1963) Bull Chem Soc Japan 30:1228
118. Zhang XM (1998) J Org Chem 63:1872
119. Parkinson CJ, Mayer PM, Radom L (1999) Theor Chem Acc 102:92
120. MacFaul PA, Wayner DDM, Ingold KU (1997) J Org Chem 62:3413
121. Fu Y, Liu L, Lin BL, Mou Y, Cheng YH, Guo QX (2003) J Org Chem 68:4657
122. Song KS, Liu L, Guo QX (2003) J Org Chem 68:262
123. Pratt DA, DiLabio GA, Valgimigli L, Pedulli GF, Ingold K (2002) J Am Chem Soc 124:11085
124. DiLabio GA, Litwinienko G, Lin S, Pratt DA, Ingold KU (2002) J Phys Chem A 106:11719
125. Tumanov VE, Denisov ET (2004) Kin Catal 45:621
126. Brinck T, Haeberlein M, Jonsson M (1997) J Am Chem Soc 119:4239
127. Wu YD, Lai DKW (1996) J Org Chem 61:7904
128. Pratt DA, de Heer MI, Mulder P, Ingold KU (2001) J Am Chem Soc 123:5518
129. Guerra M, Amorati R, Pedulli GF (2004) J Org Chem 69:5460
130. Lucarini M, Mugnaini V, Pedulli GF, Guerra MJ (2003) J Am Chem Soc 125:8318
131. Estacio SG, Cabral do Couto P, Guedes RC, Cabral BJC, Simoes JAM (2004) Theor Chem Acc 112:282
132. Lithoxoidou AT, Bakalbassis EG (2005) J Phys Chem A 109:366
133. Smith BJ, Hall NE (1997) Chem Phys Lett 279:165
134. Curtiss LA, Lucas DJ, Pople JA (1995) J Chem Phys 102:3292
135. Bach RD, Dmitrenko O (2002) J Org Chem 67:3884

Author Index Volumes 251–263

Author Index Vols. 26–50 see Vol. 50
Author Index Vols. 51–100 see Vol. 100
Author Index Vols. 101–150 see Vol. 150
Author Index Vols. 151–200 see Vol. 200
Author Index Vols. 201–250 see Vol. 250

The volume numbers are printed in italics

Subject Index

Printing: Krips bv, Meppel
Binding: Stürtz, Würzburg